T0137102

Lecture Notes in Morphogenesis

Series Editor

Alessandro Sarti, CAMS Center for Mathematics, CNRS-EHESS, Paris, France

More information about this series at http://www.springer.com/series/11247

Tamar Flash · Alain Berthoz
Editors

Space-Time Geometries for Motion and Perception in the Brain and the Arts

 Springer

Editors
Tamar Flash
Department of Computer Science
and Applied Mathematics
Weizmann Institute of Science
Rehovot, Israel

Alain Berthoz
Laboratory for Perception and Action
Collège de France
Paris, France

ISSN 2195-1934   ISSN 2195-1942   (electronic)
Lecture Notes in Morphogenesis
ISBN 978-3-030-57229-7   ISBN 978-3-030-57227-3   (eBook)
https://doi.org/10.1007/978-3-030-57227-3

This Springer imprint is published by the registered company Springer Nature Switzerland AG
The registered company address is: Gewerbestrasse 11, 6330 Cham, Switzerland

Preface

Space and time are fundamentally important for our daily living, actions, cognition, communication, social interactions, culture, and the arts. Movement—the changes in the locations and the interactions of humans with objects, environment, and living creatures—is, therefore, equally important. William James, Henri Poincaré, Albert Einstein, and many other scientists and philosophers have developed theories on how our bodies and brains shape our perception of space and time, and how this influences even our scientific ideas and theories on physical laws. Hence, understanding how the brain perceives and represents space, time, and movement, and how it plans and controls our bodily actions is of great significance. This quest requires an inter- and multidisciplinary approach linking life sciences to mathematics, computer sciences, and, last but not least, humanities and social sciences.

Combined results from behavioral and modeling studies have inspired the notion that the brain does not have a unique definition and representation of space and time, but that time depends on movement geometry. Additionally, different rules of motion and timing may be dictated by the particular mixture of Euclidean and non-Euclidian geometries being selected, depending on the action or task being carried out, the context, level of expertise, and the cognitive and emotional states of the individual. These ideas are supported by results of clinical observation by neurologists, psychiatrists, and psychologists and by child development research using brain imaging (fMRI, MEG), motion capture, and mathematical modeling studies.

These advances resonate with the statement made by William James (Pragmatism 1907): "That one Time which we all believe in and in which each event has its definite date, that one Space in which each thing has its position, these abstract notions unify the world incomparably; but in their finished shape as concepts how different they are from the loose unordered time-and-space experiences of natural men!" These notions have far-reaching implications for a variety of brain functions—the perception of biological motion, the development of motor, perceptual and cognitive skills with brain maturation and age, as well as for artistic creativity and the recognition and expression of emotion through movement in dance, music, visual arts, and film.

This book presents some of the contributions to a symposium organized by the Institute of Advanced Studies of Paris (IEA, Paris) in June 2018 in Hotel de Lauzun, Paris. The Institute has a unique Brain and Society Program directed by Prof. Gretty Mirdal for the cooperation and confrontation between social sciences and humanities with neuroscience. Tamar Flash, elected as a fellow of this Institute, has developed a project concerning the idea that artists express, each with their different approaches and methods, the fundamental properties of the mechanisms by which the brain uses geometry for perception and action. We invited mathematicians, neuroscientists, and psychologists together with artists from several disciplines to discuss this theme and to present their research and artistic works.

The chapters in this book are only a small sample of the vast potential of such an exchange. However, the new area of neuroesthetics, the use of art for the rehabilitation of pathological deficits, and the increasing interest on both sides for a creative exchange of ideas and knowledge between arts and neuroscience have encouraged us to publish this group of chapters, hoping that this will stimulate further interactions, events, and projects between the two communities. We see in dance, painting, music, theater, cinema, virtual and digital arts, etc., many possible fascinating exchanges and also applications for education, rehabilitation, and for advancing pure knowledge in the brain and cognitive sciences.

We are very grateful to Prof. Saadi Lalou, Director of the IEA, Prof. Gretty Mirdal, Dr. Simon Luck, the Scientific Director of the IEA, and all the staff of the Institute for their generous help and contributions to the scientific and practical organization of the symposium, the preparation of this book, and the great conviviality of the atmosphere in the IEA. We also wish to thank all the participants in this symposium and colleagues who contributed to the ideas that inspired this project and conference. We are also very grateful to Prof. Alessandro Sarti for offering to propose this book to Springer as part of his outstanding series.

Rehovot, Israel   Tamar Flash
Paris, France   Alain Berthoz

About This Book

This book is based on a two-day symposium at the Paris Institute of Advanced Study titled *Space-Time Geometries for Motion and Perception in the Brain and the Arts*. It includes over 20 chapters written by the leading scientists and artists who presented their related research studies at the symposium and includes six sections; the first three focus on space-time geometries in perception, action, and memory, while the last three focus on specific artistic domains: drawing and painting, dance, music, digital arts, and robotics.

There is an ever-growing interest in the topics covered by this book. Space and time are of fundamental importance for our understanding of human perception, action, memory, and cognition, and are entities that are equally important in physics, biology, neuroscience, and psychology. Highly prominent scientists and mathematicians have expressed their belief that our bodies and minds shape the ways we perceive space and time and the physical laws we formulate. Understanding how the brain perceives motion and generates bodily movements is of great significance. There is also a growing interest in studying how space, time, and movement subserve artistic creations in different artistic modalities (e.g., fine arts, digital and performing arts and music). This interest is inspired by the idea that artists make intuitive use of the principles and simplifying strategies used by the brain in movement generation and perception. Building upon a new understanding of the spatiotemporal geometries subserving movement generation and perception by the brain, we can start exploring how artists make use of such neuro-geometrical and neuro-dynamic representations in order to express artistic concepts and emotionally affect the human observers and listeners. Scientists have also started formulating new ideas of how aesthetic judgments emerge from the principles and brain mechanisms subserving motor control and motion perception.

Covering novel and multidisciplinary topics, this advanced book will be of interest to neuroscientists, behavioral scientists, artificial intelligence and robotics experts, students, and artists.

Contents

Part I
Space-Time Geometries

Chapter 1
Brain Representations of Motion Generation and Perception: Space-Time Geometries and the Arts

Tamar Flash

Abstract Many motor behavioral studies have aimed at inquiring what general principles underlie movement generation during multi-joint movements. Careful analysis of the observed behavior has led to the formulation of several kinematic laws of motion describing the invariant geometric, kinematic, and timing patterns of the upper limb and full-body movements. Similar kinematic laws of motion also characterize motion perception by the brain. Different theoretical approaches aimed at investigating the origins of these kinematic laws of motion, and the observed geometrical and temporal invariants describing motion production and perception have been developed. These included optimization theory and differential geometry models, both types of models aiming at accounting for the tight coupling existing between spatial and temporal features of natural movements. The proposed geometrical models assume that a mixture of several Euclidean and non-Euclidean space-time geometries subserve motion production and perception. This theory has enabled us also to examine the issues associated with motor compositionality—namely, how the brain constructs complex movements by composing together elementary motor primitives. Here I will also describe the possibilities offered by the unraveled motion production and perception principles for guiding research and scholarly studies concerning the expression of brain space-time geometries in movement and the arts. Open questions are how artists might be utilizing similar principles in different artistic domains, whether such principles are shared among different creative modalities, and whether they may also underlie esthetic judgments and emotional expressions in the arts.

T. Flash (✉)
Department of Computer Science and Applied Mathematics, Weizmann Institute of Science, Rehovot 76100, Israel
e-mail: tamar.flash@weizmann.ac.il

© Springer Nature Switzerland AG 2021
T. Flash and A. Berthoz (eds.), *Space-Time Geometries for Motion and Perception in the Brain and the Arts*, Lecture Notes in Morphogenesis,
https://doi.org/10.1007/978-3-030-57227-3_1

1.1 Introduction

Research in motor control focuses on the nature of the neural mechanisms subserving motion planning and control in humans and animals. In the hope of advancing our understanding, many studies have combined behavioral, theoretical, and computational models aimed at characterizing motor behavior and formulating ideas about neural control of movement. Such studies have also applied a variety of brain imaging techniques and recordings of neural activity in humans and other animals to investigate the neural representations of movement and action-perception coupling.

Over the last decade, there has also been a considerable increase in the number of behavioral and modeling studies which have recorded movements employed in different artistic domains, such as visual arts, digital arts, dance, and music. This increased interest stems from the belief that a better understanding of the neural mechanisms and brain representations of movement generation and action-perception coupling may lead to new insights into human creativity and its manifestations in various artistic expressions. Such research can throw new light on how artists take advantage of the particular spatial and temporal features of human movements and of the particularities of human perception of space and time to enhance the impact of works of art and artistic performances on human observers. The critical question is, how do artists apply their both explicit and implicit understanding of these principles and how do they benefit from such inherent knowledge to increase the emotional and intellectual impact of their creative expressions. Once we can obtain a better understanding of the particular strategies used by the brain in motion generation and perception, we may also gain a better understanding of artistic creativity.

Motion trajectories in reaching and pointing movements and in more complicated curved, obstacle avoidance, scribbling, and drawing movements display several ubiquitously observed stereotypical geometrical and temporal features. The geometric forms of drawing and curved trajectories dictate the speeds of the hand movement or of walking along curvilinear paths. In particular, movement speed varies according to the movement path's curvature and variations of curvature with time.

Another important question with respect to trajectory formation is how movement durations are selected for different movements. Behavioral studies have suggested that human movements have specific motor timing characteristics. Two kinds of temporal scaling properties referring to global and segment-wise scaling of movement durations as a function of path length have been described (see below). The observations on movement kinematics, i.e., trajectory paths, the shape of time-dependent speed profiles, and motor timing, can be accounted for by a variety of mathematical models. These models include those aimed at providing accurate descriptions of the kinematic and dynamic features of human movements and have led to the formulation of empirically observed kinematic laws of motion, such as the kinematic power laws (see below). The second type of model, for example, models based on optimal control theory, have sought explanations for the tendency of natural movements to display such stereotypical features. The third group of models, recently developed, includes those seeking to account for the observed temporal and spatial invariants of motor behavior. This type of model includes, for example, the mixture of geometries or MOG model (Bennequin et al. 2009), hypothesizing that the internal neural representations of movement are subserved by combining several Euclidean and non-Euclidean geometries (see also Handzel and Flash 1999; Flash and Handzel 2007). Research has also been conducted to identify the principles underlying motion perception. Such studies have enabled us to identify and discern similarities between the principles governing both motion production and perception of movement.

In this chapter, we first describe some of the most prominent characteristics of natural movements and the principles underlying movement planning and control. We then describe the theoretical and behavioral studies of visual perception of motion. We also report findings from several brain imaging studies investigating the brain networks subserving motion production and perception. Finally, we discuss results and ideas on space-time geometries subserving different artistic domains. We also point out potential research directions that could enlarge the scope of current research dealing with the brain and space-time geometries and movement in the arts.

1.2   Motion Planning: Kinematic Characteristics

Human hand movements display several stereotypical kinematic features visible during both planar (2D) and three-dimensional (3D) reaching and pointing actions. Starting with 2D reaching, two of the more prominent characteristics of such movements are their nearly straight geometrical paths and stereotypical bell-shaped velocity profiles of the associated hand trajectories (Morasso 1981; Abend et al. 1982; Flash and Hogan 1985). In a seminal paper, (Morasso 1981) focused on the question of whether reaching movements show greater spatial and temporal invariance when comparing end-effector (hand) trajectories to joint rotations. This simple but elegant study revealed the tendency of human subjects to generate roughly straight hand paths with symmetrical speed profiles regardless of the work-space region in which the movements were produced.

Yet, the joint angular configurations and velocity profiles vary with the end-point locations of the reaching movements, with movement direction, and the part of the work-space in which the movements are performed (see also Abend et al. 1982, Hollerbach and Flash 1982). Hence, (Morasso 1981) and several others reported that planar reaching movements display strong geometric invariance under Euclidean transformations involving translation and rotation. By contrast, the recorded joint rotations were tailored to satisfy the desired geometric and temporal features of the end-effector trajectories. This has led to the notion that multi-joint movements are planned in terms of hand trajectories rather than in terms of joint rotations. Other studies of 3D reaching (e.g., Atkeson and Hollerbach 1985) showed that during 3D reaching movements, the hands follow more curved paths, but these movements still display invariant bell-shaped tangential velocity time-dependent profiles.

Abend et al. (1982) requested subjects to generate curved movements in the horizontal plane by instructing them to avoid an obstacle or to move the hand through an

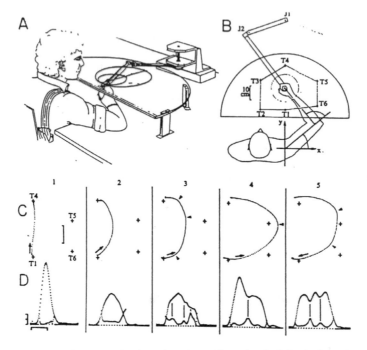

Fig. 1.1 Experimentally recorded trajectories of two-dimensional planar horizontal arm movements using a two joint manipulandum (panel A). The movements involved shoulder and elbow joint rotations (Panel B). Shown in Panel C are the hand paths, velocity, and curvature profiles. The trajectories depicted in Panel C (column 1) during point-to-point movements followed stereotypical straight hand paths and depicted symmetrical bell-shaped speed profiles. The curved trajectories (C, columns 2–5) followed curved paths with several velocity maxima and minima. The velocity minima temporally coincided with the curvature maxima (the curvature profiles are drawn below the velocity profiles). Adapted from Abend et al. (1982) and Flash and Hogan (1985)

intermediate target resulting in curved hand paths. The hand velocity profiles showed strong coupling with the hand path's curvature, revealing more complicated speed profiles with several speed maxima and minima. The hand speed minima occurred at the curvature maxima (see Fig. 1.1, panels A).

Intriguingly, similar kinematic characteristics to those reported for straight and curved hand trajectories have been reported for the trajectories of the body's center of mass during natural gait along straight and curved paths (Vieilledent et al. 2001; Hicheur et al. 2005).

1.3 Timing Characteristics

Motor timing is another aspect of hand trajectory planning that has been extensively examined in behavioral studies. Empirical studies of both point-to-point and curved upper limb movements examined how the total movement durations are modulated as a function of movement amplitude. These studies reported two ubiquitous features of motor timing and kinematics—the near invariance of movement kinematics with speed modulations and the temporal scaling of the velocity profiles with both movement size and duration. When humans perform the same motor task at different speeds, the geometrical characteristics of the movements, i.e., the path's spatial form, are nearly unaffected by the increased speed (see Hollerbach and Flash 1982; Atkeson and Hollerbach 1985; Kadmon-Harpaz et al. 2014).

Similarly, the velocity profiles of the movements had nearly the same shape, independently of the movement size. This is due to the linear scaling of movement speed by a constant scaling factor, which equals the ratios between the slow and fast movement durations. Such temporal scaling has been observed in different motor actions, including 2D and 3D reaching, curved and handwriting trajectories, locomotion, etc.

Temporal scaling is strongly associated with the isochrony principle—the phenomenon that a movement's duration is nearly independent of its amplitude. Isochrony has two different aspects. The first is the near independence of the total movement duration from its extent. This phenomenon was termed the global isochrony principle (Viviani and Schneider 1991; Viviani and Flash 1995). In contrast, local isochrony is the nearly equal durations of specific movement segments independent of their lengths. This occurs when entire trajectories, which are composed of several such consecutive parts, are globally planned (Viviani and Flash 1995).

1.4   Optimization Models

In order to account for the above kinematic characteristics of both point-to-point and curved trajectories, several mathematical models based on optimization theory have been suggested (Flash and Hogan 1985; Uno et al. 1989; Harris and Wolpert 1998; Todorov and Jordan 1998, 2002). These models are based on the assumption that the motor system optimizes particular costs, based on either kinematic or dynamic variables or neural activations.

An optimal control model successfully accounting for the kinematic characteristics of the upper limb and locomotion trajectories is the minimum jerk model (Hogan 1984; Flash and Hogan 1985). This model suggests that a primary objective of motor coordination is to generate the smoothest possible hand trajectory. This objective function was equated with the minimization of hand jerk integrated over the entire trajectory. Jerk is the rate of change of hand acceleration, and it is defined based on the third-order derivative of the hand trajectory, x(t), y(t) with respect to time, as follows:

$$\text{jerk} = \int \left(\dddot{x}^2 + \dddot{y}^2 \right) dt$$

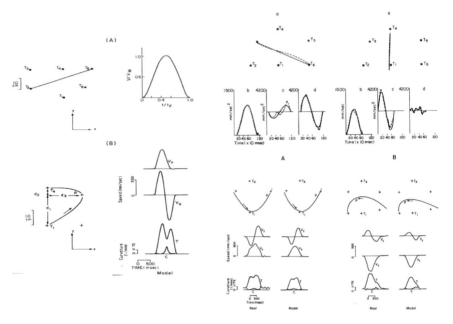

Fig. 1.2 The hand paths and velocity and curvature profiles predicted by the minimum jerk are shown in the left panel (the upper graphs display the predicted trajectories for point-to-point and curved trajectories). The lower figure displays the predicted hand paths, velocity, and curvature profiles for curved movements. In the left two figures shown are comparisons between measured (dashed lines) and predicted (solid lines) reaching movements (upper row) and curved trajectories (lower row). Adapted from Flash and Hogan (1985)

Using optimal control theory, the specific trajectories yielding the optimal performance were mathematically determined and compared to those of the measured movements (Flash and Hogan 1985). Several examples of comparisons between recorded and model-predicted trajectories for both point-to-point, i.e., reaching and curved trajectories are shown in Fig. 1.2.

Alternative models suggested that movements minimize variance (Harris and Wolpert 1998) or energetic costs or costs based on a trade-off strategy, compromising between alternative costs. One such model is the optimal feedback control model, which emphasizes the importance of optimizing feedback control and was based on an objective function representing a trade-off between accuracy and effort (Todorov and Jordan 2002). Many such studies have primarily modeled 2D reaching and curved trajectories. Only a small number of studies have dealt with 3D movements or with the inverse kinematics problem (e.g., Biess et al. 2007).

1.5 Kinematic Power Laws

A ubiquitous kinematic feature of curved hand trajectories is the strong coupling between hand speed and curvature. Not only does the hand slow down during more curved path segments, but there is also a strong dependency of the rate of slowing on the path's curvature. For example, when drawing elliptical paths, Lacquaniti et al. (1983) have reported that the hand angular velocity depends on the hand movement curvature raised by a power of $2/3^{rd}$ with a piecewise constant velocity gain factor K as follows:

$$A = K C^{2/3}$$

where A is the angular velocity and C is the Euclidean curvature: $C = \frac{\dot{x}\ddot{y} - \dot{y}\ddot{x}}{V^3}$.

The observed kinematic and temporal coupling between hand velocity and curvature is depicted in Fig. 1.3 for the drawing of elliptical trajectories. The strong coupling between geometry and speed expressed by the two-thirds power law is ubiquitous, characterizing a full spectrum of motor behaviors such as drawing (Lacquaniti et al. 1983), eye-movements (de' Sperati and Viviani 1997), whole-body locomotion (Hicheur et al. 2005, see Fig. 1.4), and speech (Tasko and Westbury 2004). This motor invariant also constrains visual perception of motion (see below and Viviani and Stucchi 1992; Levit-Binnun et al. 2006; Kandel et al. 2000; Dayan et al. 2007; Casile et al. 2010).

In order to account for the velocity modulations observed when moving along different geometrical paths, the strict two-thirds power law was later generalized (Viviani and Schneider 1991; Richardson and Flash 2002). Hence, to describe the dependency of movement speed on the geometrical form of its path, the instantaneous curvature is raised by a general exponent β (not necessarily equal to 1/3), as follows: $V = gC^{\beta}$. V marks the tangential velocity, which is related to the angular velocity and curvature, according to $V = A/C$.

The values of the exponent β providing the best match between the modeled and measured speed profiles are highly dependent on the geometrical forms of the paths (see next section).

1.6 Merging Optimization Models with Power Laws

Examining whether the alternative descriptions of hand trajectories emerged from similar organizing principles, Viviani and Flash (1995) compared the mathematical descriptions provided by the minimum jerk model to the two-thirds power-law model. Two-dimensional (2D) movements involved in drawing complex figural forms were recorded, and the of kinematic and temporal characteristics predicted by the minimum

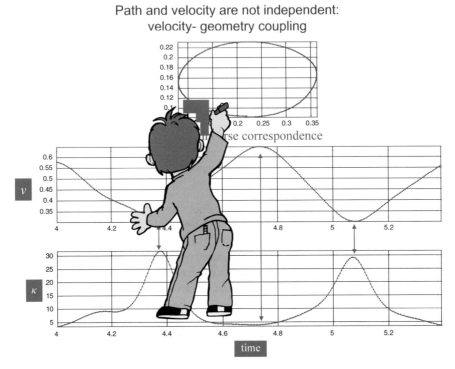

Fig. 1.3 Drawing of elliptical trajectories. Shown are examples of the hand path (upper drawing) and velocity (v) and curvature profiles (k) (lower panel). The hand velocity minima are temporally coupled with the curvature maxima, and the hand slows down during higher curvature movement segments

jerk model, the two-thirds power-law, and the isochrony principle were compared (see Fig. 1.5). The agreement between the predictions of the optimization model and the power-law description was very satisfactory. Scaling of speed within movement subunits could also be accounted for by the minimum jerk hypothesis.

Other studies have also investigated whether the two-thirds power law or similar kinematic laws emerge from the optimization of different costs. Todorov and Jordan (1998) compared the predictions of the minimum jerk model and those of the two-thirds power law and other generalized power-laws for constrained paths. The minimum jerk model provided a better description of the measured velocity profiles when moving along such paths. This study also demonstrated mathematically that the two-thirds power law is equivalent to setting the normal component of the instantaneous jerk to be zero. Other models, e.g., the minimum variance model (Harris and Wolpert 1998), have also shown that the trajectories minimizing movement variance in the presence of signal-dependent noise obey the two-thirds power law. Harris and

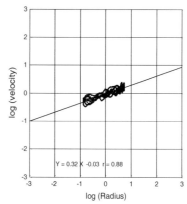

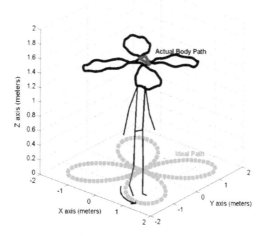

Fig. 1.4 The kinematic characteristics of the body center of mass trajectories are similar to those of recorded hand trajectories. The left panel, upper row, shows experimental measurements of gait trajectories. The right panel (top row) shows the linear relation between the logarithms of speed and the radius of curvature. The lower row shows typical cloverleaf paths recorded in these experiments. Adapted from Hicheur et al. (2005)

Wolpert demonstrated this observation through mathematical simulations of moving along a path whereby the path was constrained by having to pass through multiple via points.

Richardson and Flash (2002) used a different approach to the relation between optimization models and the power law. Assuming that movement speed complies with a generalized power law, they investigated the possibility that Mean Squared

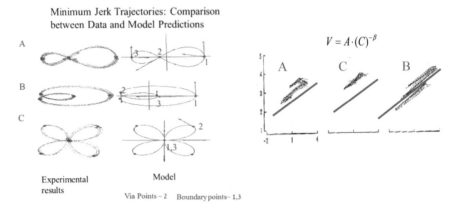

Fig. 1.5 Combining optimization and power-laws models. The left panel shows comparisons of recorded versus predicted hand paths. The predicted trajectories were predicted based on the minimum jerk model. The trajectories were predicted for three figural forms. The figures in the right panel illustrated the log (velocity) versus log (radius of curvature) behavior observed for the recorded trajectories. They were well accounted for by the minimum-jerk model. Adapted from Viviani and Flash (1995)

Derivative (MSD) optimization models can account for the empirically observed hand trajectories. MSD here is the minimization of the squared different order time-derivatives, n, of hand position. The minimum acceleration, jerk, and snap costs correspond to n = 2, n = 3, and n = 4, respectively. Enquiring what power laws are needed for the trajectory to minimize different MSD costs, perturbation theory was used to mathematically predict the beta exponent values that best minimize the integrated jerk cost for moving along complicated figural forms. Such predictions enabled to compare the values of the β exponent, calculated for the recorded movements with the beta values, predicted for different time-derivatives (n) of the trajectories used in different MSD models. A similar approach was recently taken by Huh and Sejnowski (2015) to derive the values of the β exponents for different pure spatial frequency curves. Using perturbation theory to predict what values of the exponent beta minimize jerk for various frequency curves, the derived β values were predicted to obey a general formula that expressed the dependency of the predicted exponent on the path geometry and specifically on the number of its curvature maxima and its geometrical symmetry. Quite nicely those predictions were corroborated by the empirical observations. Thus, Huh and Sejnowski (2015) demonstrated that many pure frequency curves show considerable deviations from the two-thirds power law and comply, instead with generalized power laws, where the values of the β exponents strongly depend on the geometrical form of the drawn curved path.

1.7 Geometry-Based Models

Humans interact with the environment through sensory information processing and motor actions. These interactions might possibly be understood via the underlying geometry of both perception and action. While the motor space is considered, typically, to be Euclidean, repeatable behavioral observations have suggested the possibility of other underlying geometric structures. As described below, our research hypothesizing the use of non-Euclidean geometries to describe movement kinematics has evolved from assuming that end-effector motions are represented in terms of constant equi-affine speed. This modeling approach has been further extended to a more general model involving a mixture of both Euclidean and non-Euclidean geometries.

1.8 Geometries, Groups of Transformations, and Invariants

As described by Cutting (1983), the term "invariance" was invented by several mathematicians in the nineteenth century. Following the development of this term and its spreading among mathematicians and physicists, invariance came to mean "anything which is left unaltered by a coordinate transformation." Later in the nineteenth century, with the work of Lie and Klein, the words *invariance* and *transformations* became interlocked. Cassirer (Cassirer 1938/1944, p. 19) made the strong claim that the principles of invariance and groups are the basis of both perception and geometric thought (see also, for example, Piaget 1970). According to Cassirer, Klein's geometries made it possible to bring mathematical and psychological thought together under a common denominator. Psychophysicists and psychologists have mostly learned about invariance in the context of Klein's "Erlanger Programm" of 1872. This program was set to classify and categorize the various types of geometries by the different kinds of invariance they maintain under different geometrical transformations. Thus, there is a geometrical hierarchy, as shown in Fig. 1.6. Euclidean geometry is defined by the group of rigid displacements; similarity or extended Euclidean geometry by the group of similarity transforms (rigid motions and uniform scaling); affine geometry by the group of affine transformations (arbitrary nonsingular linear mappings plus translations); and projective geometry by projective collineations.

In our current work on 2D movement generation and perception, we focus mainly on Euclidean and affine geometrical transformations and the associated motion groups (Flash and Handzel 2007; Bennequin et al. 2009) and on the equi-affine group of motions, a subgroup of affine geometry. The most constraining geometry with the largest number of invariants is Euclidean geometry associated with Euclidean transformations that preserve Euclidian distance and angles. A less strict geometry in this hierarchy is affine geometry, which preserves incidence and parallelism. Equi-affine geometry also dealt with here is a subgroup of the full affine group, which also

Nested Hierarchy of Geometries

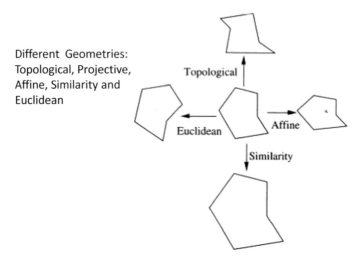

Different Geometries:
Topological, Projective,
Affine, Similarity and
Euclidean

Fig. 1.6 A nested hierarchy for Klein geometries. Depicted are the Euclidean, Similarity, affine, and topology transformations. Not depicted are the projective transformations which are situated above affine transformations

maintains area. The differential invariants of the Euclidean, full-affine, and equi-affine geometries are their canonical parametrizations (arc-lengths) and curvatures and the derivatives of different orders of the Euclidean and non-Euclidean curvatures with respect to their respective arc-lengths. For further mathematical definitions, see Faugeras (1994), Guggenheimer (1977), (Sapiro and Tannenbaum 1993), Calabi et al. (1998), Flash and Handzel (2007) and Bennequin et al. (2009).

1.9 Constant Equi-Affine Speed of 2D Trajectories and the 2/3rd Power Law

Puzzled by the coupling between velocity and curvature observed for human scribbling and drawing movements, Flash and Handzel (2007) (see also Pollick and Sapiro (1997)) searched for a more general theoretical framework, which could account for the movement phenomena described in the preceding sections. We wished to examine the nature of the geometric metrics subserving the internal representation of human movements and were interested in seeking a general framework within which one could account for the power-law and other kinematic phenomena. Starting with the behavioral observations, we have developed a mathematical framework based on differential geometry, Lie group theory, and Cartan's moving frame method for the analysis of human hand trajectories (Handzel and Flash 1999; Flash and Handzel 2007; Bennequin et al. 2009). Cartan's moving frame method (see Fig. 1.7a) enables

a

Moving Frames

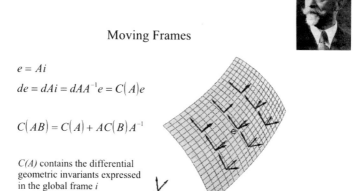

$$e = Ai$$

$$de = dAi = dAA^{-1}e = C(A)e$$

$$C(AB) = C(A) + AC(B)A^{-1}$$

$C(A)$ contains the differential geometric invariants expressed in the global frame i

b   Equi-Affine Parameterization
                                      of Curves

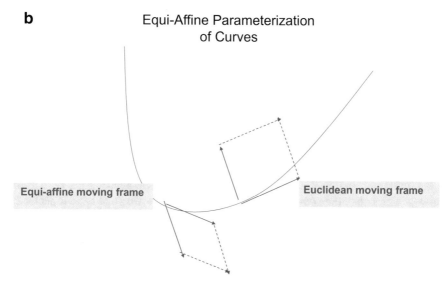

Fig. 1.7 Cartan's method of moving frames. **a** Elie Cartan (depicted here) developed a system whereby it is possible to describe an object moving along a trajectory using moving frames. This method enables to describe the differentials of the frame's unit vectors in terms of differential invariants of the geometries of the moving frames. **b** This figure depicts the difference between the Euclidean and equi-affine moving frames, enabling the use of different parametrizations of the trajectories

us to describe the trajectories of a moving object along a curve in terms of the transformations of geometrical coordinate frames and the associated differential invariants (see Cartan 1937; Flash and Handzel 2007, Bennequin et al. 2009). This line of research led us to inquire into the nature of the geometric metrics subserving the internal representation of human movements. Specifically, we demonstrated that compliance of 2D trajectories with the two-thirds power-law is mathematically equivalent to moving at a constant equi-affine speed (Handzel and Flash 1999, Flash and Handzel 2007; Pollick and Sapiro 1997). As described above, a 2D trajectory is described by the values of its time-dependent Cartesian coordinates $x(t)$, $y(t)$. The Euclidean distance along the geometrical path can be calculated based on its Euclidean metrics, ds. The Euclidean curvature corresponds to the derivative of the angular direction θ of the Euclidean velocity vector V with respect to ds. Thus, the Euclidean metric and curvature are expressed as follows:

$$ds^2 = dx^2 + dy^2, C = d\theta/ds$$

Using the equi-affine non-Euclidean geometry with its associated equi-affine moving frame (see Fig. 1.7b), the equi-affine differential form $d\sigma$ depends on the Euclidean curvature C and the Euclidean metric ds according to $d\sigma = C^{\frac{1}{3}} ds$

Equi-affine speed expresses the rate of change of the equi-affine arc-length with respect to time. Thus, for the movements to have a constant equi-affine speed, the Euclidean velocity should slow down during more curved segments, i.e., along the portions of the geometrical paths whose Euclidean curvature is larger, thus obeying the two-thirds power law.

Further studies also addressed the more general question of the origin of the symmetries of 2D motions and invariants. Polyakov et al. (2009a) investigated the issue (initially raised by Todorov and Jordan 1998) what geometric paths both minimize jerk and obey the two-thirds power law. We then showed that hand trajectories both minimize jerk and show equi-affine invariance when moving along parabolic paths.

1.10   The 1/6 Power Law for 3D Hand Trajectories

Can the two-thirds and the generalized power laws, suggested initially for 2D trajectories, be extended to 3D movements? We found that 3D hand trajectories should obey a more general power-law with the hand speed depending not only on the path's curvature but also on its torsion—expressing the rate of change of the normal to the osculating plane of movement (see Fig. 1.8). Several earlier studies had claimed that 3D drawing trajectories are piecewise planar (e.g., Soechting and Terzuolo 1987). Hence, the torsion associated with 3D hand trajectories should be nearly zero. This claim was questioned by several other studies (Pollick et al. 2009; Schaal and Sternad 2001). Extending the notion of constant equi-affine speed to the 3D case, Pollick et al. (2009) suggested that for the 3D equi-affine speed to be constant, the 1/6th power-law

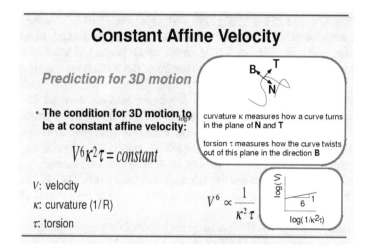

Fig. 1.8 The 1/6th power-law predicted by a constant equi-affine velocity model for three-dimensional hand trajectories. This law predicts the relationship between hand velocity, curvature, and torsion for 3D trajectories. Also shown are the Euclidean tangential (V), normal (N), and bi-normal (B) unit vectors and the linear dependency of the logarithm of the velocity (V) on the logarithms of squared curvature and torsion

must be obeyed. According to the 1/6th power-law, the hand 3D velocity v should obey the following expression:

$$v = \alpha c^{\beta} |\tau|^{\gamma},$$

where C and τ are the Euclidean curvature and torsion, respectively.

The exponent β is $-1/3$, and γ is $-1/6$. This law suggests that spatial movement speed (v) is inversely related to curvature (κ) and, to a lesser extent, to torsion (τ).

This prediction was closely corroborated in a study of 3D scribbling movements (Pollick et al. 2009). Another study closely examined a more generalized 3D power of 3D hand trajectories during the drawing of different geometrical forms (Maoz et al. 2009). The exact values of the β and γ exponents used to describe such 3D movements showed deviations from the pure 1/6th power law and considerable dependencies on the geometrical shapes of the 3D trajectories.

1.11 The Mixture of Geometries (MOG) Model

In the series of studies described above, 2D elliptical trajectories were indeed found to obey the two-thirds power law, and this law is equivalent to motion planning in terms of equi-affine geometry. However, in later studies, the two-thirds power law had to be further extended to apply to other figural forms (Viviani and Schneider

1991; Richardson and Flash 2002; Huh and Sejnowski 2015). Moreover, none of the earlier theories could successfully account for all of the spatial, kinematic, and temporal characteristics of human trajectories, and it is still unknown how the brain selects movement durations, or what the nature of the underlying motion primitives is.

Moreover, the original equi-affine description could not account for the entire spectrum of kinematic phenomena as well for the global isochrony principle. Thus, the model based on equi-affine geometric representation had to be generalized, leading to the mixture of geometries or MOG model. This model suggests that trajectory planning involves a mixture of several geometries, including Euclidian, equi-affine, and full affine geometries (Bennequin et al. 2009). The MOG model was successfully used to account for drawing and locomotion trajectories (Bennequin et al. 2009).

This extended theory was also based on the idea that movement duration is dictated by the geometry being used and that within each geometry, movement duration equals the corresponding geometric distance. Different geometries possess different canonical measures of distance along curves, i.e., different invariant arc-length parametrizations. It was therefore suggested that, depending on the selected geometry, movement duration is proportional to the corresponding arc-length parameter and that the actual movement duration reflects a particular tensorial mixture of these parameters. Near geometrical singularities, specific combinations of these parameters were assumed to be selected to compensate for time expansion or compression occurring in individual arc-length parameters. The theory was mathematically formulated using Cartan's moving frame method (Cartan 1937, Fig. 1.7a).

Assuming the mixture of geometries and the constancy of movement speed within each geometry and using Cartan's theory, the tangential velocity V was expressed as resulting from the multiplication of the affine, V_0, equi-affine, V_1, and Euclidean V_2 velocities as follows:

$$V = V_0^{\beta_0} V_1^{\beta_1} V_2^{\beta_2}$$

where

$$V_0 = C_0 k^{-\frac{1}{3}} k_1^{-\frac{1}{2}}$$
$$V_1 = C_1 k^{-1/3}$$
$$V_2 = C_2$$

where β_0, β_1 and $\beta2$ are the weight functions defined along the trajectory whose values lie within the range of [0, 1] and $\beta_0 + \beta_1 + \beta_2 = 1$. Here κ and k_1 are the Euclidean and the equi-affine curvatures, respectively, and C_0, C_1, and C_2 are constant coefficients that represent the affine, equi-affine, and Euclidean constant velocities, respectively. The predictions of the MOG model were tested on three data sets: drawings of elliptical curves, locomotion, and drawing trajectories of complex figural forms for which the ratios of movement durations for shorter versus longer loop segments were predicted. This theory succeeded in accounting for the kinematic and temporal features of the recorded locomotion and drawing movements (Fig. 1.9a, b). Comparing the predictions of the MOG model to those of the minimum jerk model, Bennequin et al. (2009), using appropriate statistical scores, showed that for human gait along curved paths, the MOG model provides more information on motor timing than the constrained minimum jerk model. For drawing, the MOG model is only slightly better than the minimum jerk model, and both models are excellent.

Hence, as argued by Bennequin et al. (2009), the principal result of the MOG model can be formulated by stating that a tensorial combination of canonical invariant

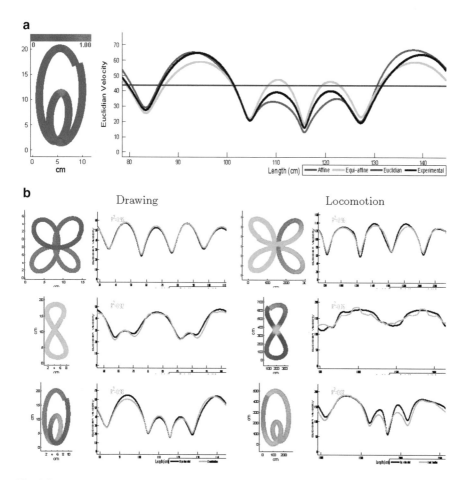

Fig. 1.9 The mixture of geometries (MOG) model. **a** Depicts the path for a double ellipse and its segmentation for a trajectory constructed using the mixture of geometries model. The panel on the right depicts the velocity profiles for the three geometrical velocities assuming constant velocities for the three geometrical parameterizations: Euclidean (blue), Equi-affine (green), and full affine (red). **b** Comparisons of measured paths and velocity profiles predicted based on the MOG model

geometric parameters gives rise to statistically non-trivial predictions not rejected by the data against which they were examined.

1.12 Motor Compositionality

The performance of any complex motor task requires the nervous system to deal with complicated cognitive, perceptual, and motor execution problems. A key idea emerging in the recent motor control literature is that most complex movements are composed of simpler elements or strokes—so-called motion primitives. These units are assumed to be combined and temporally concatenated in different ways to produce the seemingly continuous smooth movements' characteristics of human motor behavior. Different approaches and computational algorithms have been developed to infer such elementary building blocks (Flash and Hochner 2005; Abeles et al. 2013; D'Avella et al. 2015; Flash et al. 2019), but both the nature and the origins of such motion primitives are far from being understood.

One line of evidence supporting the existence of movement "letters" from which more complicated movement "words" are formed is based on the inference of discrete sub-movements derived from the decomposition of trajectories. Such sub-movements in reaching and curved movements are adequately described by uni-modular velocity profiles, well approximated by a minimum jerk (Rohrer et al. 2004) or lognormal profiles (Plamondon 1995). A reasonable assumption is that motor primitives exist at different levels of the motor hierarchy (Flash and Hochner 2005; Abeles et al. 2013; D'Avella et al. 2015). Movement primitives have been described at the level of muscle activations in the form of muscle synergies, with a large body of literature on this topic (see, for example, Overduin et al. 2012; D'avella et al. 2015). Below we briefly describe the existence of motion primitives at the level of joint kinematics, the so-called kinematic synergies. Several authors have discussed both isometric and dynamic primitives (see Hogan and Sternad 2012; Ijspeert et al. 2013).

Given the multiplicity of levels at which motion primitives can hypothetically subserve the construction of compound movements, it is quite challenging to reverse engineer the system. Among many other purposes, such a reverse engineering approach may enable us to determine why continuous motions are decomposed into smaller sub-movements in stroke and Parkinson's disease patients (Dounskaia et al. 2009). In stroke patients, following an extensive practice of reaching tasks, such sub-movements tended to blend and become less frequent, making the movements smoother and healthier-looking (Krebs et al. 1999).

An important question is whether the sub-movements revealed in these pathological conditions have a central or a peripheral origin. Possible causes could be underlying deficits in motion planning, in muscle activation, or in blending between consecutive units of action at both the kinematic and muscle activation levels. They may also reflect an intermittent mode of control or intermittent delayed feedback. These questions are currently under investigation both experimentally and theoretically.

1.13 Geometric Models Accounting for Segmentation and Compositionality

Using the minimum jerk and the equi-affine models, Polyakov et al. (2009b) examined the motor decomposition of monkeys' 2D hand trajectories by combining nerve cell recordings in motor cortex M1 and premotor cortex PMd with kinematic analysis of the scribbling movements. A Hidden Markov Model (HMM) was used to infer the neural activity states, and the identified states corresponded to parabolic segments (Polyakov et al. 2009a, b). Parabolas minimize jerk, are invariant under equi-affine transformations, and are also equi-affine geodesics. Hence, they were suggested as possible candidates for motion primitives. Polyakov et al. (2009a) further observed that following repeated practice of a target pursuit task, the number of parabolic segments decreased considerably, and the motions became smoother. In another study, Kadmon-Harpaz et al. (2019) segmented continuous monkey hand trajectories based on the identification of neural states, inferred using an HMM model.

The geometries combined in the MOG model (Euclidean, equi-affine, and affine) are each associated with their corresponding groups of transformations, i.e., group of motions. Hence, a recent study by Meirovitch (2014) (see also Flash et al. 2019) based on the MOG model presented a theory for motor decomposition. This theory assumes that motion primitives correspond to the orbits of the group of full-affine transformations. Orbits are geometric paths with constant curvatures in the associated geometries. Orbits can also be defined for the equi-affine and Euclidean groups of transformations (for further mathematical details see Meirovitch 2014; Flash et al. 2019). In Euclidean geometry, these orbits are straight lines and circles. For the equi-affine group of motions, these orbits are parabolas with zero equi-affine curvature. A hyperbola and an ellipse have constant negative and positive equi-affine curvatures, respectively (see Flash and Handzel 2007). Meirovitch (2014) used these ideas to decompose general figural forms into full-affine orbits. This model has enabled the decomposition of 2D curves into segments that both minimize jerk and are full-affine orbits. Hence, one can combine optimization and geometrical models, allowing the decomposition of human and animal trajectories into motor units.

In another study (Meirovitch et al. 2016), the minimum jerk and the MOG model were combined to deal with the speed-accuracy trade-off observed in human movements, and affine orbits were used to deal with obstacle-avoidance and on-line corrections.

1.14 Motion Perception and Neurophysiological Studies

In a pivotal study, Viviani and Stucchi (1992) demonstrated that the two-thirds power laws might apply to both motion generation and perception of motion. Subjects observed the movement of a light spot along an elliptical path and were instructed to change its motion until it appeared to move most uniformly. They did this by controlling the velocity–curvature relationships of the light spot's movement (i.e., the β exponent of the power-law equation). Subjects tended to select as most uniform motion corresponding closely to the two-thirds power law even though the dot velocity can vary by up to 200% in this type of motion. This finding was later replicated and extended using a broader range of movement speeds and elliptical paths with different eccentricities and perimeters. Those factors were found to affect the subjects' perception of motion uniformity (Levit-Binnun et al. 2006). Compatibility with the two-thirds power law was also shown to affect anticipation of perceived motion, both for handwriting movements (Kandel et al. 2000) and for simple curvilinear trajectories (Flach et al. 2004). These findings provide strong evidence that the kinematic laws of motion apply to both motion production and perception.

An fMRI study analyzed the neural correlates of kinematic constraints affecting both motion perception and production (Dayan et al. 2007). Subjects observed a dot moving along elliptical trajectories with motions that either obeyed or violated the two-thirds power law. The brain's response to motion conforming to the two-thirds power law was much stronger and more widespread than to other types of motion, including motions at a constant speed and an inverse law of motion. Compliance with the two-thirds power law was reflected in the activation of an extensive network of brain areas subserving motor production, visual motion processing, and action observation (see Fig. 1.10). These observations strongly support a central origin for this kinematic law, and that similar neural coding schemes subserve both motion perception and production.

- 1/3>Other conditions:
- Left IFG (BA 9,44,45), left PMd, left SMA and M1,
- right post-central gyrus, bilateral STS/STG, left cerebellum and left GP.
- 0 > Other conditions:
- Posterior cingulate , CCZ, fusiform gyrus, lingual gyrus, and parahippocapal gyrus. All activations were lateralized to the left.
- -1/3> other conditions:
- No activations obtained.

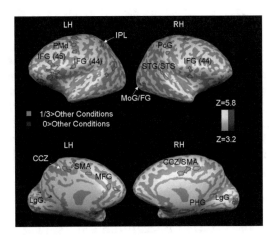

Fig. 1.10 Areas of significant activation during the perception of different types of motion. Each condition is compared with the two other conditions. CCZ, caudal cingulate zone; IFG, inferior frontal gyrus; LH, left hemisphere; LgG, lingual gyrus; MoG, middle occipital gyrus; PHG, parahippocampal gyrus; PcG, postcentral gyrus; RH, right hemisphere; SMA, supplementary motor area; STG, superior temporal gyrus; STS, superior temporal sulcus. Results are corrected for multiple comparisons at the cluster level ($P < 0.05$). Adapted from Dayan et al. (2007)

Schwartz and Moran (1999, 2000) also presented evidence in favor of the central neural origin of the two-thirds power law. They suggested that a more substantial complexity is associated with the generation of curved movements, given that this requires a continuous change in the movement direction. Such increased complexity may explain the longer reaction time needed to generate curved versus straight trajectories (Wong et al. 2016) and the longer time elapsing between initiation of neural activity in monkeys' motor cortical areas preceding movement and initiation of the movement for curved versus straight paths (Schwartz and Moran 1999, 2000).

1.15 Kinematic Power-Laws and Brain Activations in Motion Perception

Human sensitivity to biological movement was first documented by the Swedish perceptual psychologist Gunnar Johansson (1973), best known for using a point light display (PLD) as a motion stimulus. The motion of the body was displayed by attaching light bulbs to various body parts and joints. Recording various motor actions in the dark, Johansson then removed the visual information of the body by only showing the motion of white dots, corresponding to the light bulbs, against a black background. His psychophysical experiments demonstrated that human participants are able to recognize what human performers were doing through motions of the PLD. However, when the PLD was static, observers no longer recognized these actions. This phenomenon was termed biological motion perception (Blake and Shiffrar 2007). In his book chapter describing the history of his own research on biological motion perception, Cutting (2012) writes that the best rationalization of the notion of biological motion is achieved by subsuming it under the two-thirds power law. His statement should naturally be extended to the generalized power law.

The relevance of the power laws to the perception of human motion was also examined in neuroimaging studies. Calvo-Merino et al. (2006) presented dance clips and compared fMRI signals in subjects with expertise with different dance repertoires. They reported that neuronal structures in the left dorsal premotor cortex and in the ventromedial frontal cortices are selectively activated during observation of movements already in the motor repertoire in of dancers when observing these versus unpracticed movements (Calvo-Merino et al. 2005). Casile et al. (2010) examined the relevance of the two-thirds power law to the perception of human motion. In an fMRI experiment, subjects were presented with human-like avatars whose movements complied with or violated the kinematic laws of human movements. Actions complying with the two-thirds power law selectively activated specific brain areas— the left dorsal premotor cortex, dorsolateral prefrontal cortex, and medial frontal cortices. The similarities between the brain areas selectively activated in Calvo-Merino (Calvo-Merino et al. 2005; Calvo-Merino et al. 2006) and Casile et al. (2010), further suggest that these selective activations critically depend on the degree of compliance of the observed movements with normal kinematic laws of human movements. Calvo-Merino et al. (2005) also hypothesized that the activation of the ventromedial frontal cortex in the context of action observation might reflect a higher degree of either "pleasantness" or social engagement of the perceived movements.

A short-lasting attenuation of brain oscillations, called event-related desynchronization (ERD), is frequently found in the alpha and beta bands in humans during generation, observation, and imagery of movement. ERD is considered to reflect cortical motor activity and action-perception coupling. The shared information driving ERD in all these motor-related behaviors is still unknown. Recently we tested whether laws governing the production and perception of curved movement may also drive the ERD (Meirovitch et al. 2015). We characterized the spatiotemporal signature of the ERDs when human subjects observed a cloud of dots moving along elliptical trajectories, either complying with or violating the two-thirds power law. ERD in the alpha and beta bands consistently arose faster, were stronger and more widespread while observing motion obeying the two-thirds power law. An alpha activity pattern showing clear two-thirds power-law preference was observed exclusively above central motor areas. The two-thirds power law preference in the beta band was seen in additional prefrontal—central cortical sites. Compliance with the two-thirds power law is sufficient to elicit a selective ERD response in the human brain (Meirovitch et al. 2015).

We also examined the involvement of the default mode network (DMN) in the perception of biological motion (Dayan et al. 2016). The DMN appears to be involved in an array of purely cognitive and social-cognitive functions, including self-referential processing, the theory of mind, and mentalizing. More specifically, different structures within the DMN have been implicated in an array of high-level self-related functions, including self-projection, encoding, and retrieval of autobiographical memories and self-evaluation.

Although the specific internal and external stimuli that elicit DMN activity for these functions remain unknown, previous action-observation studies have suggested that the brain utilizes motion kinematics for social-cognitive processing. Using fMRI, we examined whether the DMN is sensitive to parametric manipulations of observed motion kinematics. We found that regions showing different task-induced deactivation in response to unnatural versus natural kinematics (i.e., complying versus disobeying natural power laws) of a human-like avatar largely overlapped with the DMN. The detailed analysis further revealed that unnatural human kinematics deactivated core DMN regions more strongly than natural kinematics did. These differential effects were only for the motions of realistic human-like avatars, but not for motions of an abstract visual stimulus devoid of the human form. Differences in connectivity patterns during the observation of biological versus non-biological kinematics were also observed. The DMN appears more strongly coupled with the primary nodes in the action-observation network, namely the superior temporal sulcus (STS) and the supplementary motor area (SMA), when the observed motion depicts humans rather than abstract form. These findings are the first to implicate the DMN in the perception of biological motion and may reflect the type of information used by the DMN in social-cognitive processing.

1.16   Motor Coordination

Another key topic in motor control research is the question of how the brain resolves the kinematic redundancy problem. This problem refers to the phenomena that, during most multi-joint and full-body movements, the number of degrees of freedom of the joints exceeds the number of degrees of freedom of the end-effectors (hand or foot) during the upper limb or full-body gait trajectories. Therefore, many combinations of joint rotations can achieve each foot or hand trajectory. This redundancy, therefore, raises the question of what strategies the brain uses to resolve such kinematic redundancies and how the movements of different limb segments are coordinated to achieve the desired task goals.

In recent years many research teams have focused on investigating the principles underlying motor coordination during human locomotion. A series of studies by Lacquaniti and colleagues (for review see Lacquaniti et al. 1999) have unraveled a fascinating principle of coordination among the leg segments during leg movements in full-body actions such as locomotion, running, climbing stairs, etc. Given that a leg movement in the sagittal plane involves three degrees of freedom, the question is how the brain resolves the kinematic redundancy problem of the two-dimensional mapping of trajectories of the leg's end-effector into three joint rotations of the foot, shank, and thigh. Careful empirical observations and analysis by Lacquaniti and colleagues showed that this problem is resolved through a particular pattern of intersegmental coordination—the so-called planar intersegmental law of coordination (Borghese et al. 1996; Lacquaniti et al. 1999).

This kinematic law states that during the gait cycle of human locomotion, the elevation angles of the thigh, shank, and foot do not evolve independently of each other but are coordinated to form a planar pattern of co-variation within a tilted plane (Fig. 1.11). This phenomenon has been extensively studied and found to be highly robust. The orientation of the plane describing the relations between the different leg segments is correlated with changes in gait speed and with a reduction in the expenditure of energy as the locomotion speed increases. An analytical model accounting for the observed phenomena (Barliya et al. 2009) was based on representing the movements of the thigh, shank and foot by simple oscillators for the three segments. The model required elevation angles of the leg segments to follow a simple ellipse within the intersegmental plane. This requirement resulted in specific conditions that the leg movements should satisfy. In essence, these included equal temporal frequencies of all three leg segments as well as particular constraints on the segments phase shifts. Those predictions successfully accounted for the observed coordination patterns between the different leg segments during human locomotion (Barliya et al. 2009).

Elevation angles and inter-segmental plane

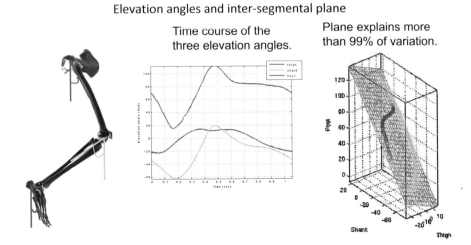

Time course of the
three elevation angles.

Plane explains more
than 99% of variation.

Fig. 1.11 The planar intersegmental coordination constraint. The elevation angles of the thigh, shank, and foot segments, shown in the left figure, are strongly correlated, and their time-dependent rotations lie on a plane. The shank and foot elevation angular rotations are almost in phase (middle panel) and get closer as speed increases. The right panel depicts the stereotypical planar "tear-drop" shape of the time-dependent configuration space vector describing the leg's modulations of the three elevation angles' rotations during a one-step cycle. Adapted from Barliya et al. 2009)

1.17 Space-Time Geometries and Compositionality in the Arts

Below we discuss several issues associated with movement generation, space-time geometries, and the arts. There are, of course, many issues and open research questions concerning art and movement generation and perception. Here, I review several studies that use the approaches discussed above to examine movement generation in different artistic domains. These studies have focused on movement compositionality, the relation between motion and emotion, and the role of the kinematic features of movements in manipulating the emotional and aesthetic impact of movements in the arts.

1.18 Drawing and Movement

Many of the aspects of motion planning and timing during curved, scribbling, and drawing movements are described above. The interested reader can also seek additional papers, for example, in a special issue of the journal Cortex on the neurobiology of drawing (Trojano et al. 2009). Here we focus only on a few issues related to drawing—the development of drawing skills, the effect of practice on drawing, and eye-hand coordination during copying and drawing.

An essential topic in drawing and movement is the development of drawing skills, which may have evolved from scribbling (Pellizzer and Zesiger 2009). With increasing perceptual-motor coordination, scribbles give rise to complex patterns, guided by visual attention and aesthetic considerations. Pellizzer and Zesiger (2009) examined the development of drawing capabilities in children by focusing on the evolvement of the dependency of speed on path geometry as captured by the two-thirds power law. An earlier study by Sciaky et al. (1987) reported that it takes several years for children to fully comply with the power law. However, babies and toddlers already show some behavior compliant with the two-thirds power law. For further interesting findings on the effect of age on compliance with the power-laws, see Viviani and Schneider (1991).

Sosnik et al. (2004) studied the effect of practice during the drawing of a sequence of movements. Subjects were required to generate drawing movements passing through a series of several static visual targets and to carry out the task as fast and as accurately as possible. Movements through any particular target set were repeated about 200 times within one day, and the subjects returned each day for eight days to perform the same tasks using several different target sets. During the first few days, the subjects tended to move through the presented targets by generating a series of straight point-to-point movements, but by the fourth day, the movements had become more curved and smoother, showing a greater co-articulation among consecutive motion segments. Furthermore, the movements, which could be divided into several trajectory chunks comprised of motions through 2–3 targets at most, showed convergence towards the behavior predicted by the minimum jerk model. Further studies of the effect of practice on drawing, including generalization across different target configurations, the impact of visual feedback and the various networks engaged during different stages of the learning of this task are described in additional studies (e.g., Sosnik et al. 2007, 2014).

Another aspect of natural drawing behavior is eye-hand coordination during both drawing and copying tasks. Miall et al. (2009) examined the nature of the cognitive and neural processes that enable us to transform visual images into a drawing. Similarly, Gielen et al. (2009) compared the eye-hand coordination strategies during both tracking and figure tracing. Tchalenko and Miall (2009) focused on the measurement of eye and hand movements of art students using several drawing and copying paradigms. Quite different eye-hand interaction strategies were observed to underlie various drawing tasks, such as copying or drawing from memory (Tchalenko and Miall 2009).

1.19 Movements and Emotion

Here we discuss how emotion affects motor coordination, as well as how film directors create and enhance the effects of different emotions in movies using camera motion. We do not present an exhaustive review of the important topic of emotion and motion, and the interested reader should refer to, e.g., De Gelder (2006), Berthoz (2003), among others.

Based on the intersegmental law of coordination and the mathematical model we developed (Barliya et al. 2009), we have studied the relationship between body expression of emotion during emotional gait and inter-joint coordination in professional actors and naive subjects (Barliya et al. 2013). Inspecting the intersegmental coordination patterns between the different leg segments, we discovered that speed is the dominant variable allowing us to discriminate among different emotional gaits. Nevertheless, the emotion expressed during normal gait can be reliably recognized based on other variables, e.g., changes in joint rotation amplitudes and modulation in the orientation of the intersegmental plane beyond those that naturally accompany merely changes in speed. Such deviations from the intersegmental pattern of coordination seen during emotionally neutral gait may reflect the effects of the emotional valence and energy expenditure during the expression of various emotions during human locomotion.

In the study by Orlandi et al. (2019), non-dancer participants were presented with contemporary dance pieces reproduced with varied or uniform acceleration and velocity ("how" the action was performed). As would be expected, the sequences characterized by considerable more timing variation were perceived as more enjoyable to observe than the same sequences uniformly executed. The former were also judged as faster, more effortful, and less reproducible than the latter. Thus, the effort also plays a vital role in the aesthetic experience of dancing. Moreover, the perceived effort does not just relate to the kinds of movements being performed, but also to their kinematic complexity. Camurri et al. (2016) also developed computational models aimed at studying the effects of different dance qualities on the aesthetic experience of the observers.

We have also used optic flow analysis to assess the effect of local and global motion on the emotions elicited in people watching movies (Dayan et al. 2018). Film theorists and practitioners have noted that motion can be manipulated in movie scenes to supplement its narrative in eliciting emotional responses in viewers. However, we still have only a limited understanding of the role of movement in generating an emotional percept. On the one hand, mirroring real-life perceptual processing, movies continuously depict *local motion*—the movement of objects, including humans, crucial for generating an emotional response. Movie scenes also frequently portray *global motion*, mainly induced by large camera movements.

We used fMRI to elucidate the contribution of local and global motion to emotion perception under the rich and unconstrained visual settings associated with movie viewing. Brain activity in areas showing preferential responses to emotional rather than to neutral content was strongly linked over time with frame-wide variation in global motion fields, and to a lesser extent, with local motion information. Since global motion fields are experienced during self-motion, our results suggested that camera movement might induce illusory self-motion cues in viewers, which interact with the movie's content in generating an emotional response. Overall, these findings strongly implicated motion signals in the perception of emotion. Many issues remain open concerning on how movie directors benefit from using different mixtures of space-times geometries to achieve the cognitive and emotional effects on film observers they wish to achieve. Some of these issues, especially concerning the existence of four sensorimotor and perceptual spaces, are discussed in the chapters by Berthoz, Granier, and Bennequin included in this volume, and in Bennequin and Berthoz (2017).

1.20   Conclusions and Future Directions

The behavioral, modeling, and brain mapping studies discussed here can be applied and generalized to characterize the spatial and temporal features of movements used in different artistic creations, such as dance, music performance and conducting, and in the fine arts. Similar approaches may also be employed to assess motion perception in different artistic domains. Such approaches may enable us to quantitatively evaluate the emotional responses elicited in human observers in response to a variety of creative works and styles. Further research can also examine whether aesthetic judgments of artistic works rely on the kinematic laws of motion and the temporal invariants unraveled during everyday motor behavior. Specifically, it is not yet clear whether the manipulation of some of the reported movements' spatial, temporal, and kinetic features is indeed used in a variety of artistic domains. Similarly, one could examine what temporal or geometrical deformations of natural behavior can cause works of art to appear more or less engaging or to lead to aversive emotional responses.

Focusing on motion capturing of dancers', musicians', or painters' movements to compare their kinematic and temporal regularizes to those observed for motor production of more mundane actions would allow us to examine whether and how these regularities are maintained or manipulated in the arts to generate different esthetic and emotional responses in human observers. Several attributes that affect aesthetic judgments and emotional percepts when listening to music or watching dance, involve temporal features such as tempo, rhythm, duration of different portions of the dance or music themes, etc. The MOG model has enabled us to associate different geometrical and temporal attributes of movement and, therefore, may guide the development of new ideas of how space and time are related in different artistic modalities. In a series of studies examining the movements involved in the generation of Graffiti, Berio et al. (2020) have examined style in drawing by applying ideas and models building upon current models of motion compositionality and optimization. Alternatively, there may exist some particular levels of abstraction at which the same principles subserve artistic production in a variety of creative modalities, such as music, dance, painting and drawing, and digital arts. Such research directions may inspire further studies focusing on space-time geometries, motion generation, and perception and how such attributes may subserve artistic creativity and expression.

Here we mainly focused on movement kinematics and on spatial and temporal invariants characterizing natural motor behavior. Several important topics, not discussed here, include movement dynamics, energetics, and motor variability both within individual subjects and across different individuals. Motor variability, dynamics, and energetics have important roles in enabling artists to express their individual creative ideas, traits, and emotions. Motor variability, which is naturally associated with the performance of even the simplest motor behaviors, lends artists with vast opportunities for expanding the spectrum and versatility of their motor repertoire and artistic works of art. This, in turn, can greatly expand the richness and impact of their artistic and emotional expression. Although not addressed here, these topics are of great interest concerning motion production and perception in the arts. Hence, the approaches described here to inquire into the nature of space-time geometries, and motor compositionality can significantly contribute to such future research and scholastic studies.

Acknowledgements I am grateful to many of my former and current students and postdocs for their highly valuable contributions to the studies described in this chapter. I greatly value the contributions of Amir Handzel, Magnus Richardson, Felix Polyakov, Uri Maoz, Armin Biess, Ido Bright, Eran Dayan, Ronit Fuchs, Yaron Meirovitch, Avi Barliya, Irit Sella, Matan Karklinsky, and Naama Kadmon-Harpaz and David Ungarish for their significant contributions. I also am particularly grateful to Daniel Bennequin and Alain Berthoz for their collaboration in the research on neural space-time geometries for motion and perception. I also acknowledge the valuable contributions of Neville Hogan, Paolo Viviani, Frank Pollick, Moshe Abeles, Emilio Bizzi, Martin Giese, Antonio Casile, and Rivka Inzelberg to the different studies described in the chapter and to Jenny Kien for her editorial help. The studies described in this chapter received support from our earlier EU funded Amarsi, Vere, and Tango projects, and more recently, from the CNCRS BSF-NSF grant. The experimental work was conducted at the Moross Laboratory for Vision Research and Robotics at the Weizmann Institute. Tamar Flash was, until recently, the incumbent of the Dr. Hymie Moross professorial chair. The research was also supported by research grants from the Estate of Naomi K. Shapiro, the Rudolph and Hilda U. Forchheimer Foundation, the Consolidated Anti-Aging Foundation, the Steven Gordon Family Foundation, and the Levine Family Foundation.

References

Abeles, M., Diesmann, M., Flash, T., Geisel, T., Hermann, M., & Teicher, M. (2013). Compositionality in neural control: An interdisciplinary study of scribbling movements in primates. *Frontiers in computational neuroscience, 7,* 103.

Abend, W., Bizzi, E., & Morasso, P. (1982). Human arm trajectory formation. *Brain: a Journal of Neurology, 105*(Pt 2), 331–348.

Atkeson, C. G., & Hollerbach, J. M. (1985). Kinematic features of unrestrained vertical arm movements. *Journal of Neuroscience, 5*(9), 2318–2330.

Barliya, A., Omlor, L., Giese, Ma., & Flash, T. (2009). An analytical formulation of the law of intersegmental coordination during human locomotion. *Experimental Brain Research, 193,* 371–385.

Barliya, A., Omlor, L., Giese, Ma., Berthoz, A., & Flash, T. (2013). Expression of emotion in the kinematics of locomotion. *Experimental Brain Research, 225,* 159–176.

Bennequin, D., & Berthoz, A. (2017). Several geometries for the generation of movement. In Laumond et al. Geometric and Numerical Foundations of Movements. Springer.

Bennequin, D., Fuchs, R., Berthoz, A., & Flash, T. (2009). Movement timing and invariance arise from several geometries. *PLoS Computational Biology, 5,* e1000426.

Berio, D., Leymarie, F. F., & Calinon, S. (2020). Interactive Generation of Calligraphic Trajectories from Gaussian Mixtures. In *Mixture Models and Applications* (pp. 23–38). Springer: Cham.

Berthoz, A. (2003). *Emotion and Reason.* The Cognitive Science of Decision Making: Oxford University Press, Oxford.

Biess, A., Liebermann, D. G., & Flash, T. (2007). A computational model for redundant human three-dimensional pointing movements: Integration of independent spatial and temporal motor plans simplifies movement dynamics. *Journal of Neuroscience, 27*(48), 13045–13064.

Blake, R., & Shiffrar, M. (2007). Perception of human motion. *Annual Review of Psychology, 58,* 47–73.

Borghese, N. A., Bianchi, L., & Lacquaniti, F. (1996). Kinematic determinants of human locomotion. *Journal of Physiology, 494,* 863–879.

Calabi, E., Olver, P. J., Shakiban, C., Tannenbaum, A., & Haker, S. (1998). Differential and numerically invariant signature curves applied to object recognition. *International Journal of Computer Vision, 26*(2), 107–135.

Calvo-Merino, B., Glaser, D. E., Grèzes, J., Passingham, R. E., & Haggard, P. (2005). Action observation and acquired motor skills: An FMRI study with expert dancers. *Cerebral Cortex N. Y. N, 1991*(15), 1243–1249.

Calvo-Merino, B., Gre'zes, J., Glaser, D. E., Passingham, R. E., & Haggard, P. (2006). Seeing or doing? Influence of visual and motor familiarity in action observation. *Current Biology, 16,* 1905–1910.

Camurri, A., Volpe, G., Piana, S., Mancini, M., Niewiadomski, R., Ferrari, N., & Canepa, C. (2016, July). The dancer in the eye: towards a multi-layered computational framework of qualities in movement. In *Proceedings of the 3rd International Symposium on Movement and Computing* (pp. 1–7).

Cartan, E. (1937). La The´orie des Groupes Finis et Continus et la Ge´ome´trie Diffe´rentielle Traite´es par la Me´thode du Repe're Mobile. Gauthier-Villars, Paris: Cahiers Scientifiques, vol. 18.

Casile, A., Dayan, E., Caggiano, V., Hendler, T., Flash, T., & Giese, Ma. (2010). Neuronal encoding of human kinematic invariants during action observation. *Cerebral Cortex, 20,* 1647–1655.

Cassirer, E. (1938). The concept of group and the theory of perception. (A. Gurwitsch, trans.). *Philosophy and Phenomenological Research, 1944, 5,* 1–35. (Originally published, 1938).

Cutting, J. E. (1983). Four assumptions about invariance in perception. *Journal of Experimental Psychology: Human Perception and Performance, 9*(2), 310–317.

Cutting, J. E. (2012). Gunnar Johansson, events, and biological motion. In K. Johnson & M. Shiffrar (Eds.), *People watching: Social, perceptual, and neurophysiological studies of body perception (Chapter 2).* New York: Oxford University Press.

D'avella, A., Giese, M., Ivanenko, Y. P., Schack, T., & Flash, T. (2015). Editorial: Modularity in Motor Control: from Muscle Synergies to Cognitive Action Representation. *Frontiers in Computational Neuroscience, 9.*

Dayan, E., Casile, A., Levit-Binnun, N., Giese, M. A., Hendler, T., & Flash, T. (2007). Neural representations of kinematic laws of motion: evidence for action-perception coupling. *Proceedings of the National Academy of Sciences, 104,* 20582–20587.

Dayan, E., Sella, I., Mukovskiy, A., Douek, Y, Giese, M. A., Malach, R., & Flash, T. (2016). The default mode network differentiates biological from non-biological motion. *Cerebral Cortex (New York, n.y.:1991). 26,* 234–245.

Dayan, E., Barliya, A., de Gelder, B., Hendler, T., Malach, R., & Flash, T. (2018). Motion cues modulate responses to emotion in movies. *Scientific Reports, 8*(1), 1–10.

De Gelder, B. (2006). Towards the neurobiology of emotional body language. *Nature Reviews Neuroscience, 7*(3), 242–249.

De' Sperati, C., & Viviani, P. (1997). The relationship between curvature and velocity in two-dimensional smooth pursuit eye movements. *The Journal of Neuroscience: The Official Journal of the Society for Neuroscience (J Neurosci), 17,* 3932–3945.

Dounskaia, N., Fradet, L., Lee, G., Leis, B. C., & Adler, C. H. (2009). Submovements during pointing movements in Parkinson's disease. *Experimental Brain Research, 193*(4), 529–544.

Faugeras, O. (1994). Cartan's moving frame method and its application to the geometry and evolution of curves in the Euclidean, affine and projective planes. In: Mundy, J. L., Zisserman, A., Forsyth, D., (eds.), Applications of Invariance in Computer Vision, New York: Springer, vol. 825. pp. 11–46.

Flach, R., Knoblich, G., & Prinz, W. (2004). The two-thirds power law in motion perception. *Visual Cognition, 11*(4), 461–481.

Flash, T., & Handzel, A. A. (2007). Affine differential geometry analysis of human arm movements. *Biological Cybernetics, 96,* 577–601.

Flash, T., & Hochner, B. (2005). Motor primitives in vertebrates and invertebrates. *Current Opinion in Neurobiology, 15,* 660–666.

Flash, T., & Hogan, N. (1985). The coordination of arm movements: an experimentally confirmed mathematical model. *Journal of Neuroscience, 5,* 1688–1703.

Flash, T., Karklinsky, M., Fuchs, R., Berthoz, A., Bennequin, D., & Meirovitch, Y. (2019). Motor compositionality and timing: Combined geometrical and optimization approaches. In *Biomechanics of Anthropomorphic Systems* (pp. 155–184). Springer: Cham.

Gielen, C. C., Dijkstra, T. M., Roozen, I. J., & Welten, J. (2009). Coordination of gaze and hand movements for tracking and tracing in 3D. *cortex, 45*(3), 340–355.

Guggenheimer, H. W. (1977). Differential Geometry. Dover Publications.

Handzel, A. A., & Flash, T. (1999). Geometric methods in the study of human motor control. *Cognitive Studies, 6*(3), 309–321.

Harris, C. M., & Wolpert, D. M. (1998). Signal-dependent noise determines motor planning. *Nature, 394,* 780–784.

Hicheur, H., Flash, T., & Berthoz, A. (2005). Velocity and curvature in human locomotion along complex curved paths: a comparison with hand movements. *Experimental Brain Research, 162,* 145–154.

Hogan, N. (1984). An organizing principle for a class of voluntary movements. *Journal of Neuroscience, 4*(11), 2745–2754.

Hogan, N., & Sternad, D. (2012). Dynamic primitives of motor behavior. *Biological Cybernetics, 106*(11–12), 727–739.

Hollerbach, J. M., & Flash, T. (1982). Dynamic interactions between limb segments during planar arm movement. *Biological Cybernetics, 44*(1), 67–77.

Huh, D., & Sejnowski, T. J. (2015). A spectrum of power laws for curved hand movements. *Proceedings of the National Academy of Sciences, 112*(29), E3950–E3958.

Ijspeert, A. J., Nakanishi, J., Hoffmann, H., Pastor, P., & Schaal, S. (2013). Dynamical movement primitives: Learning attractor models for motor behaviors. *Neural computation, 25*(2), 328–373.

Johansson, G. (1973). Visual perception of biological motion and a model for its analysis. *Perception & Psychophysics, 14,* 201–211.

Kadmon-Harpaz, N., Flash, T., & Dinstein, I. (2014). Scale-invariant movement encoding in the human motor system. *Neuron, 81,* 452–462.

Kadmon-Harpaz, N., Ungarish, D., Hatsopoulos, N. G., & Flash, T. (2019). Movement decomposition in the primary motor cortex. *Cerebral Cortex, 29*(4), 1619–1633.

Kandel, S., Orliaguet, J. P., & Viviani, P. (2000). Perceptual anticipation in handwriting: the role of implicit motor competence. *Perception and Psychophysics, 62,* 706–716.

Krebs, H. I., Aisen, M. L., Volpe, B. T., & Hogan, N. (1999). Quantization of continuous arm movements in humans with brain injury. *Proceedings of the National Academy of Sciences of the United States of America, 96,* 4645–4649.

Lacquaniti, F., Terzuolo, C., & Viviani, P. (1983). The law relating the kinematic and figural aspects of drawing movements. *Acta Psychologica (Amst.), 54,* 115–130.

Lacquaniti, F., Grasso, R., & Zago, M. (1999). Motor patterns in walking. *Physiology, 14*(4), 168–174.

Levit-Binnun, N., Schechtman, E., & Flash, T. (2006). On the similarities between the perception and production of elliptical trajectories. *Experimental Brain Research, 172,* 533–555.

Maoz, U., Berthoz, A., & Flash, T. (2009). Complex unconstrained three-dimensional hand movement and constant equi-affine speed. *Journal of Neurophysiology, 101,* 1002–1015.

Meirovitch, Y. (2014). Movement decomposition and compositionality based on geometric and kinematic principles. Department of Computer Science and Applied Mathematics., Ph.D. dissertation, Weizmann Inst. Sci., Rehovot, Israel.

Meirovitch, Y., Harris, H., Dayan, E., Arieli, A., & Flash, T. (2015). Alpha and beta band event-related desynchronization reflects kinematic regularities. *Journal of Neuroscience, 35,* 1627–1637.

Meirovitch, Y., Bennequin, D., & Flash, T. (2016). Geometrical invariance and smoothness maximization for task-space movement generation. *IEEE Transactions on Robotics, 32*(4), 837–853.

Miall, R. C., Gowen, E., & Tchalenko, J. (2009). Drawing cartoon faces–a functional imaging study of the cognitive neuroscience of drawing. *Cortex, 45*(3), 394–406.

Morasso, P. (1981). Spatial control of arm movements. *Experimental Brain Research, 42*(2), 223–227.

Orlandi, A., Cross, E. S., & Orgs, G. (2019). Timing is everything: Aesthetic perception of movement kinematics in dance. Cognition, 205, 104446.

Overduin, S. A., d'Avella, A., Carmena, J. M., & Bizzi, E. (2012). Microstimulation activates a handful of muscle synergies. *Neuron, 76*(6), 1071–1077. https://doi.org/10.1016/j.neuron.2012.10.018.

Pellizzer, G., & Zesiger, P. (2009). Hypothesis regarding the transformation of the intended direction of movement during the production of graphic trajectories: A study of drawing movements in 8-to 12-year-old children. *Cortex, 45*(3), 356–367.

Piaget, J. (1970). *Structuralism.* New York: Basic Books.

Plamondon, R. (1995). A kinematic theory of rapid human movements. *Biological Cybernetics, 72*(4), 295–307.

Pollick, F. E., & Sapiro, G. (1997). Constant affine velocity predicts the 1/3 power law of planar motion perception and generation. *Vision Research, 37*(3), 347–353.

Pollick, Fe, Maoz, U., Handzel, Aa, Giblin, Pj, Sapiro, G., & Flash, T. (2009). Three-dimensional arm movements at constant equi-affine speed. *Cortex., 45,* 325–339.

Polyakov, F., Stark, E., Drori, R., Abeles, M., & Flash, T. (2009a). Parabolic movement primitives and cortical states: merging optimality with geometric invariance. *Biological Cybernetics, 100,* 159–184.

Polyakov, F., Drori, R., Ben-Shaul, Y., Abeles, M., & Flash, T. (2009b). A compact representation of drawing movements with sequences of parabolic primitives. *PLoS Computational Biology, 5,* e1000427.

Richardson, M. J., & Flash, T. (2002). Comparing smooth arm movements with the two-thirds power law and the related segmented-control hypothesis. *Journal of Neuroscience, 22*(18), 8201–8211.

Rohrer, B., Fasoli, S., Krebs, H. I., Volpe, B., Frontera, W. R., Stein, J., et al. (2004). Submovements grow larger, fewer, and more blended during stroke recovery. *Motor Control, 8*(4), 472–483.

Sapiro, G., & Tannenbaum, A. (1993). Affine invariant scale-space. *International Journal of Computer Vision, 11*(1), 25–44.

Schaal, S., & Sternad, D. (2001). Origins and violations of the 2/3 power law in rhythmic three-dimensional arm movements. *Experimental Brain Research, 136,* 60–72.

Schwartz, A. B., & Moran, D. W. (1999). Motor cortical activity during drawing movements: Population representation during lemniscate tracing. *Journal of Neurophysiology, 82,* 2705–2718.

Schwartz, A. B., & Moran, D. W. (2000). Arm trajectory and representation of movement processing in cortical motor activity. *European Journal of Neuroscience, 12*(6), 1851–1856.

Sciaky, R., Lacquaniti, F., Terzuolo, C., & Soechting, J. (1987). A note on the kinematics of drawing movements in children. *Journal of Motor Behavior, 19,* 518–525.

Soechting, J., & Terzuolo, C. (1987). Organization of arm movements in three-dimensional space: Wrist motion is piecewise planar. *Neuroscience, 23,* 53–61.

Sosnik, R., Hauptmann, B., Karni, A., & Flash, T. (2004). When practice leads to co-articulation: the evolution of geometrically defined movement primitives. *Experimental Brain Research, 156,* 422–438.

Sosnik, R., Flash, T., Hauptmann, B., & Karni, A. (2007). The acquisition and implementation of the smoothness maximization motion strategy are dependent on spatial accuracy demands. *Experimental Brain Research, 176*(2), 311–331.

Sosnik, R., Flash, T., Sterkin, A., Hauptmann, B., & Karni, A. (2014). The activity in the contralateral primary motor cortex, dorsal premotor, and the supplementary motor area is modulated by performance gains. *Rontiers in Human Neuroscience, 8,* 201.

Tasko, S. M., & Westbury, J. R. (2004). Speed–curvature relations for speech-related articulatory movement. *Journal of Phonetics, 32,* 65–80.

Tchalenko, J., & Miall, R. C. (2009). Eye-hand strategies in copying complex lines. *Cortex, 45*(3), 368–376.

Todorov, E., & Jordan, M. I. (1998). Smoothness maximization along a predefined path, accurately predicts the speed profiles of complex arm movements. *Journal of Neurophysiology, 80,* 696–714.

Todorov, E., & Jordan, M. I. (2002). Optimal feedback control as a theory of motor coordination. *Nature Neuroscience, 5,* 1226–1235.

Trojano, L., Grossi, D., & Flash, T. (2009). Cognitive neuroscience of drawing: contributions of neuropsychological, experimental, and neurofunctional studies. *Cortex, 45,* 269–277.

Uno, Y., Kawato, M., & Suzuki, R. (1989). Formation and control of optimal trajectory in human multijoint arm movement. *Biological Cybernetics, 61*(2), 89–101.

Vieilledent, S., Kerlirzin, Y., Dalbera, S., & Berthoz, A. (2001). Relationship between velocity and curvature of a human locomotor trajectory. *Neuroscience Letters, 305*(1), 65–69.

Viviani, P., & Flash, T. (1995). Minimum-jerk, two-thirds power law, and isochrony: converging approaches to movement planning. *Journal of Experimental Psychology: Human Perception and Performance, 21*(1), 32.

Viviani, P., & Schneider, R. (1991). A developmental study of the relationship between geometry and kinematics in drawing movements. *Journal of Experimental Psychology: Human Perception and Performance, 17*(1), 198–218.

Viviani, P., & Stucchi, N. (1992). Biological movements look uniform: evidence of motor-perceptual interactions. *Journal of Experimental Psychology: Human Perception and Performance, 18*(3), 603.

Wong, A. L., Goldsmith, J., Krakauer, J. W. (2016). A motor planning stage represents the shape of upcoming movement trajectories. *Journal of Neurophysiology, 116,* 296–305.

Chapter 2
A Common Multiplicity of Action Spaces in the Brain and in the arts? The 4/5 Spaces Theory

Alain Berthoz

Abstract The main idea of this chapter is to propose a theory that suggests that the brain has different networks for different action spaces. The five spaces which are considered are (1) *Body space* (BS); (2) *Peri-personal* space (PPS), or reaching, or prehension space, often called «*near space*»; (3) *Extrapersonal space* (EPS), sometimes called «*far space*»; (4) *Far environmental space* (FES) in which we «navigate»; and (5) Imaginal space (IS). This modularity has been suggested by neuropsychological and neurological pathologies. Recent studies using brain imaging support the existence of different brain networks subserving these different action spaces and a specific review of the literature is done here for some of these spaces. Theoretical work from Daniel Bennequin and Tamar Flash support the possibility that different geometries are implemented in the brain to meet the different processes that are necessary for action in these spaces. In addition, these geometries may be subclasses of a more general geometry (Topos), as described in the chapter of Daniel Bennequin. This would allow both specialization of these networks and compatibility allowing an efficient transition from one to another. It is possible that, during development, the brain of children implements these geometries to allow manipulation of reference frames and perspective changes also in cognitive functions. This theory leads to a new interpretation of psychiatric and neurological pathologies.

While space is perceived as unitary, experimental evidence indicates that the brain actually contains a modular representation of space, specific cortical regions being involved in the processing of extra-personal space, that is the space that is far away from the subject, and that cannot be directly acted upon by the body, while other cortical regions process peripersonal space, that is the space that directly surrounds us and which we can act upon.[1]

---

[1]Cléry (2015).

A. Berthoz (✉)
Collège de France, CIRB, Paris, France
e-mail: alain.berthoz@college-de-france.fr

© Springer Nature Switzerland AG 2021
T. Flash and A. Berthoz (eds.), *Space-Time Geometries for Motion and Perception in the Brain and the Arts*, Lecture Notes in Morphogenesis,
https://doi.org/10.1007/978-3-030-57227-3_2

2.1 Introduction[2]

Evolution has offered the brain a challenge. Living organisms have to move, feed, compete, reproduce in the physical space of the planet. This challenge required that living forms would deal with the multiplicity of spaces in which they live in order to survive. For them the planet contains forms, and offers challenges, whose size and distance range from microscopic to environmental scales, from molecular, or even submolecular, to a few meters, from proteins conformational changes to thousands of miles traveled by birds, fishes and whales for reproduction. In addition the first humans travelled long distances by foot from Africa. Their survival, in this great diversity of *action spaces* could not have been secured without some simplifying principles which I have called «simplex».[3] The necessity of flexibility to change from one scale of space to others, in an ever changing world, also had to be taken care of and flexible boundaries had also to be established by specific neural mechanisms, at least in complex organisms like us.

I will present here a theory which suggest that this complexity have been solved, for the case of movement, navigation, by the «simplex» choice of organising space with a small number of principles and a modular organisation of brain networks. These principles are based upon a segregation of brain modules which work in *different geometries* (We have recently given a formal theory of these different geometries as can be seen in the chapter of Daniel Bennequin in this book). These geometries have to be different because they corresponds to different actions spaces:

1. *Body space* (BS)
2. *Peri-personal* space (PPS), or reaching, or prehension space, often called «*near space*». The size of this space is within arm length.
3. *Extrapersonal space* (EPS)[4] which is «beyond reaching».[5] Unfortunately it has often called «*far space*» in many studies or immediate locomotor space. The size of this space is a few meters. Reaching a target, in this space, requires to stand and walk a few steps.
4. *Far environmental space* (FES) in which we «navigate».

One could add also a fifth one, imaginal space (IS), which will not be considered here. The *time scales* and distances of these action spaces are very different. Here we cannot consider all these different scales and we shall mention only a few of the mechanisms involved in this modularity. Very few empiricial studies have suggested such dissociation.[6]

---

[2]I would like to state that because this book arose from a meeting in the Institute of Advanced Studies in Paris this paper will be very speculative as the idea of such Institute is to stimulate us to get «out of our confort zone» and risk new hypothesis. I have been priviledged to be also very much inspired by the works and pioneering ideas of Tamar Flash and Daniel Bennequin on this question.

[3]A. Berthoz. Simplexity. Harvard Univ; Press 2016

[4]Marco et al. (2019).

[5]Flanders et al. (1999).

[6]Josephs and Konkle (2019).

I shall first review some recent neurophysiological and clinical evidence in favor of this modularity. Then I will try to show that artists have understood this segregation and are revealing the differents spaces, their different geometries, and their boundaries. An example of this modularity, inspired by our theory, is given by the chapter of François Garnier in this book.

2.2 Why is Action Important for Understanding the Modularity of Different Spaces?

2.2.1 Action at the Foundation of Geometry

The idea that action it at the foundation of geometry and of the concept of space is not new. I will here only quote two statements by eminent mathematiciens (My traduction from french). Henri Poincaré wrote[7] «To localise a point in space is simply to imagine the movement necessary to reach it It is not a question of representing the movements themselves but simply the muscular sensations which accompagny them». He also insited upon the idea that the notion of space arises from the fact that the ideintity of objects stems from our movements relative to them[8]. There is no a priori space, but exploitation of a set of transformations, described by groups. Albert Einstein wrote[9]: «Poincaré is right …The fatal error that a mental necessity, preceding all experience, is at the basis of Euclidian geometry is due to the fact that the empirical basis on which is based the axiomatic construction of euclidian geometry was forgotten. Geometry must be considered as a physical science whose utility must be judged by its relation with "*l'expérience sensible*"». Philosophers like Maurice Merleau-Ponty and Henri Bergson also gave a fundamental role to action in the foundation of our perception of space.

Even Physicist J. C. Maxwell layed down an embodied "epistemology of muscular effort". He wrote "Some minds can go on contemplating with satisfaction pure quantities presented to the eye by symbols, and to the mind in a form which none but mathematicians (or platonician linguists) can conceive. Others, ….calculate the forces with which the heavenly bodies pull at one another, and they feel their own muscles straining with the effort. To such men momentum, energy, mass are not mere abstract expressions of the results of scientific inquiry. They are words of power, which stir their souls like the memories of childhood".[10] I suppose he would have said the same for the concept of space and agree with Poincaré and Eisntein.

---

[7] Poincaré (1907).

[8] See Daniel Bennequin chapter and publications and the publicatiosn of Kevin O' Reegan.

[9] Albert Einstein. Conceptions scientifiques. Flammarion, p. 29.

[10] Address to the Mathematical and Physical Sections of the British Association, Liverpool, Sept. 15, 1870.

2.2.2 Non Euclidian Geometries in the Brain

In other chapters of this book Tamar Flash and Daniel Bennequin have given evidence
that the brain does not use Euclidian geometries, and that affine and equi-afine geometries underly both gestures, locomotion and visual perception.[11] This departure of
perceived space from Euclidian was suggested also by the pioneering works of Jan
Koenderink.[12] He wrote that: «Optical space differs from physical space. The structure of optical space has generally been assumed to be metrical. In contradiction,
we do not assume any metric, but only incidence relations (i.e., we assume that
optical points and lines exist and that two points define a unique line, and two lines a
unique point)...The condition that makes such an incidence structure into a projective
space is the Pappus condition. The Pappus condition describes a projective relation
between three collinear triples of points, whose validity can-in principle-be verified empirically. The Pappus condition is a necessary condition for optical space
to be a homogeneous space (Lobatchevski hyperbolic or Riemann elliptic space) as
assumed by, for example, the well-known Luneburg theory. Apparently optical space
is not totally different from a homogeneous space, although it is in no way close to
Euclidean». He also suggested that the curvature changes from elliptic in near space
to hyperbolic in far space. At very large distances the plane becomes parabolic. The
suggestion that visual space is an affine transformation of physical space has been
challenged.[13] Finally he studied the fact that when we look at a painting representing
a portrait the face seems always to keep looking at us irrespective of our position and
viewing angle. From this he concluded that «...the psychogenesis of visual awareness maintains a number—at least two, but most likely more—of distinct spatial
frameworks simultaneously involving "cue–scission." Cues may be effective in one
of these spatial frameworks but ineffective or functionally different in other ones.».[14]

Recently, with Daniel Bennequin, we have attempted to combine the fact that
the brain uses different networks for different geometries with the theory that these
geometries are not only Euclidian. Daniel Bennequin has propose that the *different
action spaces* would use geometries that would be sub-classes of a more general type
of geometry: the geometry of Topos.[15] We are presently trying to obtain empirical
evidence supporting this theory. Many examples are available of diffferent geometries
in the brain in relation with action. For example, we have shown that in the superior
colliculus a remarkable change is performed in the geometry in which visual space
is coded. Wether the geometry of the retina is a classical spheric geometry in which
each point in space is projected to a spherical coordinate system, in the superior
colliculus the coordinates are log–polar. Whe have shown that this geometry prepares

---

[11] See also: Bennequin (2009).

[12] Koenderink et al. (2002).

[13] Wagner et al. (2018).

[14] Koenderink et al. (2016).

[15] Bennequin and Berthoz (2017).

the transformation of the retinotopic coding of the visual world into the space of the eye muscles motor system reference frame.[16]

2.3 The Distinction Between Dorsal/«Near» and Ventral/«Far» Action Spaces

2.3.1 The Categorisation from Neurology

The proof of a modularity between different actions spaces come from several sources. The first one is neurology. It has been known since the early days of neurology that different lesions in the brain subserved deficits in respectively what was known as personal (body) space, peri-personal, and extra-personal space. One can find in the neurological literature, for instance in the papers by O. J Grüsser (Fig. 2.1) the categorisation of *grasping space, instrumental grasping space* (which can be extended by the use of tools[17]), *near distant action space, far distant action space* and *visual background*. As shown in Fig. 2.1, grasping space can itself be divided into whole body space, manual, peri-oral and intra-oral spaces.

Very recently a very interesting set of observations has been made by a clinical psychologist Chantal Lheureux who identified in autistic children that they had a division of space with boundaries («*bord*» in french).[18] She identified that autistic children were often limited to one of the four spaces described above and had difficulties tranfering their activities to antother. Often they kept close to «borders» like walls of virtual limits of these difference spaces. It would be intersting to relate this behaviour to the discoveries of «border cells» in the hippocampus[19] and, in the field of painting or other arts like dance or theater, to see how these borders, or boundaries, are defined. However these boundary cells belong to the Euclidian/Topological geometry for navigation, they don't limit the space of a geometry with respect to another one, they limit a piece of space inside a given geometry.

2.3.2 Lateralisation

It seems that there is a lateralisation for the treatment of near and far space. For example the right cerebral hemisphere has been considered to be specialized for spatial attention and orienting. We cannot here review the vast knowledge of the

---

[16]Tabareau N., Bennequin D., Berthoz A., Slotine J. J., Girard B. Geometry of the superior colliculus mapping and efficient oculomotor computation. *Biol. Cybern.* 97(4): 279–292. See also in my book "Simplexity" op.cit. p 146

[17]Forsberg et al. (2019).

[18]Lheureux (2018).

[19]Stewart et al. (2013); Moser et al. (2017).

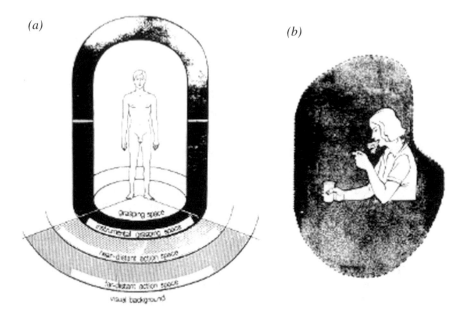

Fig. 2.1 The different action spaces as identified by neurologists. **a** Grasping space, instrumental grasping space, near distant action space, far distant action space, visual background. The outer borders of the near-distant action space use about 6–8 meters away from the subject. The grasping space can be extended by using instruments. **b** The grasping space is subdivided into subcompartments the general (whole body), manual perioral and intraoral grasping space are illustrated in this figure (Adapted From O. Grüsser and T. Landis), Visual agnosias and other disturbances of visual perception and cognition. Cronly-Dillon J.R. (Ed.), Vision and ... de I Rentschler - 1994).

right/left brain different functions. It is well known that the left brain mostly deals with language and with details, or objects, and the right brain is more involved in global evaluation of the world. We have shown that during navigation in space the left brain is more involved in the sequential egocentric operations and the right brain more in the allocentric treatment.[20] The right hemisphere seesm to plays a dominant role in the processing of space close to the body.[21] This dominance is reduced at farther distances, whether hand motor actions are involved or not. Longo et al. [22] have suggested that the right occipito-temporal cortex may be specialized not just for the orientation of spatial attention generally, but specifically for orienting attention in the near space immediately surrounding the body. It has been suggested[23] that *right ventral occipital cortex is involved in far-space search*, and right frontal eye field is involved regardless of the distance to the array. It was found that *right posterior parietal cortex* is involved in search only in *far space*, with a neglect-like effect when

---

[20]Khonsari et al. (2007); Ghaem et al. (1997); Igloi et al. (2010, 2014); Lambrey et al. (2007, 2011).

[21]Lucas Rinaldi op. cit.

[22]Longo et al. (2015); Coello et al. (2003).

[23]Laeng et al. (2002); Mahayana et al. (2014).

the target was located in the most eccentric locations. No effects were seen for any site for a feature search task.

An interesting question is: how does the brain coordinates the shifts between the different brain modules involved in near and far space? It has been suggested that a common eye centering mechanism[24] could be used, or even that the vestibular system may be a common reference frame for different spaces. Another view proposed by Daniel Bennequin (See his chapter) is that these different spaces would be sub-classes on the more general geometry of Topos and because they would be linked through this geometry they would be fundamentally compatible reducing the complexity of the shifts between them.

2.3.3 Boundaries Between Near and Far Spaces Are Flexible.

A large literature in monkeys and humans does show that the prehension space can be extended by tool use.[25] This was elegantly shown by Atsushi Iriki[26] in a series of experiments in the mondey. He showed that the receptive field of tactile neurons of fingers in the cortex can be extended away to the tip of a tool uses to grasp an object Recently it has even been shown in humans that we can also perceive the tactile properties of the grasped or touched distant object with the tool.[27] But this extension of peripersonal prehension space cannot be considerd as a critical argument against the idea that there are different geometries for near and far space. On the contrary the fact that the space extends with the use of a tool show that action is the essence of the problem and therefore the same geometry can, and has, to be uses as far as the brain will have to produce an action of the body of the same nature. It has even been shown[28] that the presence of others or avatars in virtual reality experiments and social interactions may extend the extend of «near» space or even bring the far space closer.[29] By contrast people judge a given geographical distance as subjectively smaller when they can exert control across that distance.[30] This has, of course to be related with the more general problem of «proxemy», the optimal distance for social interactions.[31] Similarly to adults, the boundary between near and far space is not

---

[24]Pieter Medendorp and Douglas Crawford (2002).

[25]Brozzoli et al. (2010); Costantini (2014).

[26]Maravita and Iriki (2004); Obayashi et al. (2001); Bonifazi et al. (2007); Alessandro Farne et al., The Role Played by Tool-Use and Tool-Length on the Plastic Elongation of Peri-Hand Space: A Single Case Study. Cogn Neuropsychol, 22(3), 408–418.

[27]Miller (2018).

[28]Fini et al. (2014); Griffiths and Tipper (2012); Near or Far? It Depends on My Impression: Iachiniet al. (2015).

[29]Fini et al. (2015).

[30]Wakslak and Kyu Kim (2015).

[31]Perry et al. (2016).

fixed in children and both active tool use and verbal labels can modulate this uncertain boundary.[32]

2.4   Cognitive Strategies for «Far Environmental» Space

We deal here with far environmental navigation space (FES) like travelling in an appartment, or a city, or outdoors. It constitutes the *third space of my theory of space representation in paitings*. In paintings it corresponds to the background of the scene. (Mountaines, cities, room walls in indoors pictures, the sea in marine paintings etc.). I shall here only give a very brief account of the main networks involved in the strategies whih are used by the brain for navigating in these spaces or for memory of travelled paths, in order to show their modularity and the potential differences of the geometries involved.

2.4.1   Egocentric, Allocentric, Heterocentric Reference Frames.

The brain has different networks for different cognitive strategies of navigation and spatial memory of travelled routes:

(1)   It can use a first person perception of the world. The «near» peripersonal prehension space (PPS) (catch a glass) and «far» beyond reach extrapersonal space (EPS) (stand up and go to the door) described above can both involve a «fist person view point» (1PP), also called *egocentric visuo-spatial reference frame, or strategy.* It is mainly implemented in the left hippocampus and right cerebellum.[33] This strategy is also used for planning, or memorizing, sequential navigation of trajectories in the «far environmental space» (FES) (go to the post office). It can also be involved when observing a picture.

(2)   The brain can take another person or an object in the environment and use a third person, or *heterocentric*, perspective (3PP).

(3)   The brain can use an *allocentric* perspective (cartographic) independent of any personal point of view (FES) (Mainly involving the right hippocampus and left neo-cerebellum).[34]

One of the challenges has been, and is still, to understand how we shift from one strategy to the other. This requires not only modification of the coordinates (as occur

---

[32]Scorolli (2016).

[33]Blouin et al. (1993).

[34]These two strategies already exist in insects like honeybees et therefore have been a very early appearing process in evolution (Menzel et al. 2000). See also the pinoeering paper: Woodin and Allport (1998).

in the change, mentionned above, of retinotoppic coding in the retina to the log-polar coding in the superior colliculus for instance,) coordinate shifts but also viewpoint change, i.e. perspective change.

2.4.2 Perspective Change.

Perspective is presented as a late discovery of the Italian renaissance but in fact it can be found much earlier in Asian arts for instance. The multiplication of perspectives has been used by visual arts over the civilisations for a very long time even before the formal theory of Euclidian perspective was proposed in Italy. However I will leave to the vision experts to discuss this point. One question which arose in neuroscience has concerned the mechanisms which allow the brain to *shift from one perspective to the other*. This requires some kind of changes of geometrical coordinate system but more generally some deep restructuring of the perceived space and the relations with our own body. The same kind of operation probably occurs in front of a painting when we examin the different perspectives which have been chosen by the artist either explicitly as in Picasso, or implicitly as in the famous painting «Les Ménines» by Diégo Velasquez.

We have discussed above the problem of shifting from near to the so called «far» space in the current literature. Specific work should be done to understand how the brain operates transitions between these spaces. We know more about the perspective changes occuring between egocentric and allocentric perspectives.

Although several networks have been identified depending upon the context and specific conditions in which this transition occurs[35] a common set of area have been identified which constitute a core network of the perspective changes.[36] Several cortical areas are involved (Dorsal pariétal, precuneus, retrosplenial, parieto-temporal, parieto-occipital) which are involved in encoding and retrieving target locations from different perspectives and are modulated by the amount of viewpoint rotation. The parahippocampus is also involved in this process as it can code environmental landmarks. It is activated during shift of ego to allocentric viewing of a scene as demonstrated by intracranial recording of field potential activity in epileptic patients.[37]

The *retrosplenial cortex* received a particular attention. It was suggested that it is involved in coordinate transformations from ego to allo centric coding.[38] It has also been suggested that it is involved in anticipating an upcoming perspective change during a priming paradigm in which the future perspective is cued in advance.[39]. It is also important to note that the retrosplenial cortex is part of the Papez circuit

---

[35]Schmidt et al. (2007).

[36]Sulpizio et al. (2013).

[37]Bastin et al. (2012).

[38]Epstein et al. (2005); Bicanski and Burgess (2016); Lambrey (2012).

[39]Sulpizio et al.

involved in the regulation of emotions, and therefore may also be involved in the emotional evaluation of visual scenes.[40]

It may not be always necessary to shift perspective, for example from ego to allocentric reference. In fact some paintings offer this combined perspective point of view. For example the brain can store a visual scene in both egocentric and allocentric perspective, with an *oblique view* adopted recently by many virtual reality softwares for navigation. We have described the brain activities involved in either egocentric, allocentric and oblique (or slanted) view perspectives.[41]

One should therefore be careful not to try to attribute a rigid set of brain networks to the general question of geometries. It seems that the brain can build configurations of networks adapted to each particular conditions of interaction with the world and this may be the same for art. For instance Laure Rondi-Reig and her team have studied the networks involed in *exploration* versus *exploitation* of a paths in space during navigation.[42] The identified networks do contains the main areaas mentioned above and obey to a princple of laterlisation (noet the interesting correspondance between hippocampus and the contralateral cerebellum). But the networks are very different when the subject navigates with these two type of behaviour.

2.5   Representing Space in Paintings

2.5.1   The Three Spaces Theory

The way artists represent space has been a very important subject for art historians but it is not my aim here to review this literature which is not in my field of competence. A number of recent studies have addressed the specific question of geometries in art works.[43] I would like to suggest that many artists and particularily painters have understood the modularity and segregation of the different action spaces described above. They organise their paintings accordingly in depth. The scene is often divided in at least three planes corresponding resptively to: (1) reaching or immediate periper-sonal space, (PPS); (2) Extrapersonal or near distant locomotor action space (EPS); and (3) environmental far distant action space (FES). In french these are often called: (1) *premier plan;* (2) *deuxième plan;* (3) *fond.*

Generally, if some humans or animals, grounds, stones, stairs, or trees are shown in the closest (PPS) plane their size is rather great and occupies a sizable potion of the total height of the picture. Trees often occupy the whole hight of the picture as is also seen in Japanese or Chinese paintings. Humans or animals located in the second space have generally a smaller size which is about half of the size of the persons in

---

[40]Sulpizio et al. (2015b).

[41]Barra et al. (2012).

[42]Babayan et al. (2017).

[43]Baldwin et al. (2014); Burleigh et al. (2018).

the first space. Finally the background show buildings or natural scenes the size is again a proportion of the two first spaces.

The contrast betwen these three spaces is most of the time increased by the clearness, and often dark colour, for near space, or fuzziness, and often light blue, for far space. Coulour is also used to distinguish these spaces. For example the foreground may be dominated by red colours as red is perceived by the brain as stands up in the forefront (Political women tend to use red dresses when photographed in the midst of groups or males!!), and red traffic lights used for signals on the road meaning «stop» or forbidden. The painter Jean-Baptiste Corot has used a very specific trick to separate spaces by often adding in the middle of the painting a red dot, often placed on the clothes of a person in the second space, hence helping the painting to acquire this depth separation. Sometimes these separation in three spaces is not clearly perceptible because the painter has used methods to link these spaces. For example sometimes a path (often curved and set in a diagonal of the picture space) links the three spaces and creates a depth organisation which transcends the separation between the three action spaces. But a carefull examination of the painting shows that they are clearly present. Objects in far space can be brought into the brain's near-space through tool-use. A near object can be pushed into far space by changing the pictorial context in which it occurs and this shift may be related to the different roles of the dorsal and ventral pathways in the brain.[44]

But it is also true that sometimes artists negate this division and tend to fuse the three spaces. It would however be interesting to study these paintings in detail to see if this apparent fusion does not reveal also some clear distinction. Another method used by artist is the produce a gradient of depth by varying the size of the objects, trees, persons, buidlings in a continuous way in depth. A special method for linking the spaces which has been used by many painters, including Czanne and imprressionists, is to paint a path or road which goes obliquely from the foreground to the background.

It should be understood that I am not trying to say that this separation of the different action spaces, corresponding to different geometries in the brain, is clearly made in all paintings. What I am saying is that is does appear in many paintings, even sometimes hidden by transverse perspective cues. I submit that artist have understood the modularity of basic brain mechanisms and do use this modularity as a way may be to interact more directly with the viewer. A consequence of my theory is also that some artists who have understood this mode of operation of the brain break it on purpose in order to create the surprise and disconfort or discongruence which is one of the aims of Art.

---

[44]Nicholls et al. (2011).

2.5.2 Right/Left Asymetry

The second interesting link between geometrical feature of paintings and brain modularity is the lateralisation of both brain and paintings which has been discussed above. A *right-left asymetry* is obvious in most paintings. I have the feeling that, generally, the near reference objects, walls, or trees or even persons, are drawn on the right of the paintings. A beautifull example is the disposition of most of the «Annunciations» paintings of the Italian Renaissance which has been studied by Sarah Longo.[45] The angel generally comes from the far space on the the left, corresponding to the right brain, and the virgin which is on the right. She has a repertoire of action gestures (surprise, denial, acceptation and submission etc..) and is clealy in the space where the body is present and «touchable» by the viewer. Between the symbolic «far» space of the angel and very far environmental and imaginary space of God (on the left), and the action and terrestrial «near», and very far space of the virgin (on the right), a column marks often the separation between the two spaces and, I believe, the differential treatment between the two brains (left brain and right visual space for near, and right brain and left visual space for far). One can also see, in most of the annunciations, the separation in depth of the near and far and environmental spaces. Perspective links all these modularity in an apparently coherent organisation and, often, a column marks the boundary betwen the spaces.

2.5.3 Challenges

I have tried here to describe how acting in space is dealt with by the brain and the modularity underlying several action spaces. We have seen that several types of modularity are present as announced in the introduction:

- Distinct networks for «*near*» (peripersonal reaching space PPS) and «*far*» (extrapersonal out of reach space EPS). *Boundaries* between these near and far spaces are *flexible* and action, tool use, or context, or social factors, or emotion dependant.
- Distinct networks are involved in *egocentered, allocentered, heterocentered, object centered* cognitive strategies for spatial memory and navigation.
- The *left and right brain* are involved in distinct functions in these processes.
- Distinct networks are involved in *exploration* and *exploitation* of spatial environmental tasks. The equivalent in the interaction with a painting for instance could be the notion of *familiarity* (when we see a painting for the first time we explore and when we have seen it many trimes we exploit).

---

[45]Giuseppe Longo, Sara Longo. Infini de dieu et espaces des hommes en peinture, condition de possibilité pour la révolution scientifique: "Les mathématiques dans l'oeuvre d'art", ISTE-WILEY, 2020. Version largement amplifiée et revue d'un texte paru dans ISTE OpenScience—Published by ISTE Ltd, London,UK, 2019, et, en forme très préliminaire en français, dans "Le formalisme en action : aspects mathématiques etphilosophiques", (J. Benoist, T. Paul eds.) Hermann, 2013.

- Finally *gender differences*[46] between men and women strategies and performance exists during spatial tasks. They involve neurohormonal factors (oestrogenes and testosterone).

Specific networks are involved in *perspectives changes* but even these networks are dependant upon action, context etc.. It is obvious that when producing or perceiving a painting or any art work our brain does involves a variety of these mechanisms in parallel or in sequence. No general theory allows us to understand how these diferent approaches to space are coordinated by our brain. Artists are able to show us empirically all these different components of our perception and play with them. The richness of the repertoire of mechanism si also the root of artistic infinite creativity, of artistic vicariance.[47]

The theory proposed here concerns only one very specific aspect of the organisation of space in pictural Arts. In a painting at least three geometrical systems are combined: (a) perspective; (b) the three (or four) depth spaces mentionne above; (c) left/right asymetry. But many more can be, and are, used. For example Maurice Merleau-Ponty had an original view of depth perception. He thought that perception of depth is produced by a change in view point by which we perceive depth as «a width seen as a profile».[48] He wrote: «to consider a depth as a width seen as a profile, a subject has to leave his place, his point of view on the world, and thinks about himself in a sort of ubiquity» (p. 294). This process is different from the idea discussed about, and illustrated in Fig. 2.1 of several spaces embeded in a series of sourrounding spaces. It would actually be interesting to study of this has been used by artists. Lastly the impact of brain strokes on painting has been explored[49] and it may be interesting to review this literature from the point of view proposed in this chapter. Namely try to see if in the post–stroke paintings of the patients one can identify a specific deficit in the spatial distinctions proposed above.

References

Babayan, B. M., et al. (2017). A hippocampo-cerebellar centred network for the learning and execution of sequence-based navigation. *Science Report, 7*(1), 17812.

Baldwin, J., et al. (2014). Comparing artistic and geometrical perspective depictions of space in the visual field. *Iperception, 5*(6), 536–547.

Barra, J., et al. (2012). Does an oblique/slanted perspective during virtual navigation engage both egocentric and allocentric brain strategies? *PLoS ONE, 7*(11), e49537.

Bastin, J., et al. (2012). Timing of posterior parahippocampal gyrus activity reveals multiple scene processing stages. *Human Brain Mapping.* https://doi.org/10.1002/hbm.21515

---

[46]Grön et al. (2000); Lambrey and Berthoz (2007); Perrochon et al. (2018); Lauer et al. (2019). Also See the books by Doreen Kimura. «Sex and cognition», and by Melissa Haines. «Brain gender».

[47]Berthoz (2013). La vicariance. Le cerveau créateur de monde O. Jacob. Paris 2013 La Vicarianza. Il nostro cervello creatore del mondi. Eudice. Milan 2013.

[48]«Une largeur vue de profil».

[49]Petcu et al. (2016).

Bennequin D., & Berthoz, A. (2017). Several geometries for the generation of movement. In Laumond et al. (Ed.), *Geometric and numerical foundations of movements*. Springer.

Bennequin, D., Fuchs, R., & Flash, A. B. T. (2009). Movement timing and invariance arise from several geometries. *PLoS Computing Biology, 5*(7), e1000426.

Berthoz, A. (2013). The vicarious brain. Creator of worlds. Yale University Press.

Bicanski, A., & Burgess, N. (2016). Environmental anchoring of head direction in a computational model of retrosplenial cortex. *Journal of Neuroscience, 36*(46), 1601–11618.

Blouin, J., et al. (1993). Reference systems for coding spatial information in normal subjects and a deafferented patient. *Experimental Brain Research, 93*(2), 324–331.

Bonifazi, S., et al. (2007). Dynamic size-change of peri-hand space through tool-use: Spatial extension or shift of the multi-sensory area. *Journal of Neuropsychology, 1*(1), 101–14.

Brozzoli, C., et al. (2014). Action-specific remapping of peripersonal space. *Neuropsychologia, 48*(3), 796–802.

Burleigh, A., et al. (2018) Natural perspective: Mapping visual space with art and science. *Vision (Basel), 2*(2), 21.

Cléry, J. (2015). Neuronal bases of peripersonal and extrapersonal spaces, their plasticity and their dynamics: knowns and unknowns. *Neuropsychologia, 70,* 313–326.

Coello, Y., et al. (2003). Vision for spatial perception and vision for action: A dissociation between the left-right and near-far dimensions. *Neuropsychologia, 41*(5), 622–633.

Costantini, M. (2014). When a laser pen becomes a stick: Remapping of space by tool-use observation in hemispatial neglect. *Experimental Brain Research, 232*(10), 3233–3241.

Di Marco, S., et al. (2019). Walking-related locomotion is facilitated by the perception of distant targets in the extrapersonal space. *Scientific Reports, 9*(1), 9884.

Epstein R. A., et al. (2005). Learning places from views: Variation in scene processing as a fucntion of experience and navigational ability. *Journal of Cognitive Neuroscience 17,* 73–83.

Fini, C., et al. (2014). Sharing space : The presence of other bodies extends the space judged as near. *PLoS One, 9*(12), e114719.

Fini, C., et al. (2015). Social scaling of extrapersonal space: target objects are judged as closer when the reference frame is a human agent with available movement potentialities. *Cognition, 134,* 50–56.

Flanders, M., et al. (1999). Reaching beyond reach. *Experimental Brain Research, 126*(1), 19–30.

Forsberg, A., et al. (2019). Tool use modulates early stages of visuo-tactile integration in far space: evidence from event-related potentials. *Biological Psychology, 145,* 42–54.

Ghaem, O., et al. (1997). Mental navigation along memorized routes activates the hippocampus, precuneus and insula. *NeuroReport , 8,* 739–744.

Griffiths, D., & Tipper, S. P. (2012). When far becomes near: Shared environments activate action simulation. *Quarterly Journal of Experimental Psychology (Hove), 65*(7), 1241–1249.

Grön, G., et al. (2000). Brain activation during human navigation: Gender-different neural networks as substrate of performance. *Natural Neuroscience, 3*(4), 404–408.

Iachini, M. T., et al. (2015). Information and spatial behavior in virtual interactions. *Acta Psychol (Amst), 161,* 131–136.

Igloi, K., et al. (2010). Lateralized human hippocampal activity predicts navigation based on sequence or place memory. *Proceedings of National Academy Sciece USA, 107*(32), 14466–14471.

Iglói, K., Doeller, C. F., Paradis, A. L., Benchenane, K., Berthoz, A., Burgess, N., & Rondi-Reig, L. (2014) Interaction between hippocampus and cerebellum crus I in sequence-based but not place-based navigation. *Cereb Cortex. 25*(11), 4146–4154.

Josephs, E. L., & Konkle, T. (2019). Perceptual dissociations among views of objects, scenes, and reachable spaces. *Journal of Experimental Psychology: Human Perception and Performance, 45*(6), 715–728. https://doi.org/10.1037/xhp0000626

Khonsari, et al. (2007). Lateralized parietal activity during decision and preparation of saccades. *Neuroreport, 18*(17), 1797–1800.

Koenderink, J. J., et al. (2002). Pappus in optical space. *Percept Psychophys, 64*(3), 380–391.

Koenderink, J., et al. (2016). Facing the spectator. *Iperception, 7*(6), 2041669516675181.

Laeng, B., et al. (2002). Multiple reference frames in neglect ? An investigation of the object-centred frame and the dissociation between "near" and "far" from the body by use of a mirror. *Cortex, 38*(4), 511–528.

Lambrey, S., et al. (2007). Distinct visual perspective-taking strategies involve the left and right medial temporal lobe structures differently. *Brain, 131*(pt2), 523–534.

Lambrey, S., et al. (2011). Imagining being somewhere else: Neural basis of changing perspective in space. N.*Cereb Cortex*.

Lambrey, S. (2012). Imagining being somewhere else: Neural basis of changing perspective in space. *Cereb Cortex, 22*(1), 166–174.

Lambrey, S., & Berthoz, A. (2007). Gender differences in the use of external landmarks versus spatial representations updated by self-motion. *Journal of Intergrative Neuroscience, 6*(3):379–401.

Lauer, J. E., et al. (2019). The development of gender differences in spatial reasoning: A meta-analytic review. *Psychology Bulletin, 145*(6), 537–565.

Lheureux, C. (2018). Angoisses spatiales et création de l'espace dans la clinique de l'autisme. David-seERESI«Le Coq-héron» 2018/4 N° 235 | pages 50 à 60ISSN 0335-7899ISBN 9782749262208. https://www.cairn.info/revue-le-coq-heron-2018-4-page-50.htm.

Longo, M. R., et al. (2015). Right hemisphere control of visuospatial attention in near space. *Neuropsychologia, 70*, 350–357.

Mahayana, I. T., et al. (2014). Far-space neglect in conjunction but not feature search following transcranial magnetic stimulation over right posterior parietal cortex. *Journal of Neurophysiol, 111*(4), 705–714.

Maravita, A., & Iriki, A. (2004). Tools for the body (schema). *Trends in Cognitive Science 8*(2), 79–86.

Menzel R., et al. (2000). Two spatial memories for honeybee navigation. *Proceedings of Biology Science, 267*(1447), 961–968.

Miller, L. E. (2018). Sensing with tools extends somatosensory processing beyond the body. *Nature, 561*(7722), 239–242.

Moser, E. I., et al. (2017). Spatial representation in the hippocampal formation: A history. *Natural Neuroscience, 20*(11), 1448–1464.

Nicholls, M. E. R., et al. (2011). Near, yet so far: The effect of pictorial cues on spatial attention. *Brain Cognitive, 76*(3), 349–352.

Obayashi, S., et al. (2001). Functional brain mapping of monkey tool use. *Neuroimage, 14*(4), 853–861.

Perrochon, A., et al. (2018). The influence of age in women in visuo-spatial memory in reaching and navigation tasks with and without landmarks. *Neuroscience Letters, 25*(684), 13–17.

Perry, A., et al. (2016). The role of the orbitofrontal cortex in regulation of interpersonal space: Evidence from frontal lesion and frontotemporal dementia patients. *Social Cognitive and Affective Neuroscience, 11*(12), 1894–1901.

Petcu, E. B., et al. (2016). Artistic skills recovery and compensation in visual artists after stroke. *Frontier Neurology, 7*, (76).

Pieter Medendorp, W., & Douglas Crawford, J. (2002). Visuospatial updating of reaching targets in near and far space neuroreport, *13*(5), 633–636.

Poincaré, H. (1907). La Science et l'hypothèse.

Schmidt, D., et al. (2007). Visuospatial working memory and changes of the point of view in 3D space. *NeuroImage, 36*(3), 955–968.

Scorolli, C. (2016). Reaching for objects or asking for them: Distance estimation in 7- to 15-year-old children. *Journal of Motor Behavior, 48*(2), 183–191

Stewart, S., et al. (2013). Boundary coding in the rat subiculum. *Philosophical Transactions of the Royal Society B, 369*(1635), 20120514.

Sulpizio, V., et al. (2015b). Visuospatial transformations and personality: Evidence of a relationship between visuospatial perspective taking and self-reported emotional empathy. *Experimental Brain Research, 233*(7), 2091–2102.

Sulpizio, V., et al. (2013). Selective role of lingual/parahippocampal gyrus and retrosplenial complex in spatial memory across viewpoint changes relative to the environmental reference frame. *Behavior Brain, 242,* 62–75.

Wagner, M., et al. (2018). Differentiating between affine and perspective-based models for the geometry of visual space based on judgments of the interior angles of squares. *Vision (Basel), 2*(2).

Wakslak, C. J., & Kyu Kim, B. (2015). Controllable objects seem closer. *Journal of Experimental Psychology General, 144(3),* 522–527.

Woodin, M. E., & Allport, A. (1998). Independent reference frames in human spatial memory: Body-centered and environment-centered coding in near and far space. *Memory and Cognition, 26*(6), 1109–1116.

Chapter 3
Variety of Brains Geometries
for Action/Perception

Daniel Bennequin

Abstract In this text we present a collection of various inventions in Geometry, from Euclid to Grothendieck and Thom, passing by Kepler, Euler, Riemann and Poincaré, and we indicate for all of them a deep link with the organization and functions of the brains of animals, especially humans. In particular, we present works conducted with Tamar Flash and Alain Berthoz, and many other collaborators, using a variety of geometries, all included in the affine geometry, which are combined to prepare, plan and control the execution of human movements, for drawing or for walking. We also present a recent model, elaborated with David Rudrauf, Kenneth Williford, and other collaborators, showing that the three-dimensional projective geometry underlines a space for consciousness. Additionally, we introduce a new kind of geometry developed with Alain Berthoz, and many collaborators, which is able to organize different action spaces (at least five); this geometry corresponds to a 2-category of fibered categories of geometrical spaces over several networks in the brain, seen as different sites for different topos. The unifying theme of these works is adaptation.

3.1 Euclid and Poincaré

Euclidian geometry (the "elements" were written around -300) has a reputation of fixity. However, as shown by Poincaré ($+1900$, cf. Poincaré 1902), this particular geometry describes the form of our exchanges with the external world in time. The space all around us, which seems so familiar, is still mysterious; Poincaré asked explicitly the question: what is the nature of this space? (Fig. 3.1).

He suggested an answer: discriminating between voluntary actions and the environmental changes is vital for any living entity. Therefore an internal mechanism allows the comparison between active and passive transformations of our sensory data. The residual ambiguity between these two kinds of transformations takes the

D. Bennequin (✉)
University Paris 7 Mathematics Place Aurélie Nemours, 75013 Paris, France
e-mail: bennequin@math.univ-paris-diderot.fr

© Springer Nature Switzerland AG 2021
T. Flash and A. Berthoz (eds.), *Space-Time Geometries for Motion and Perception in the Brain and the Arts*, Lecture Notes in Morphogenesis,
https://doi.org/10.1007/978-3-030-57227-3_3

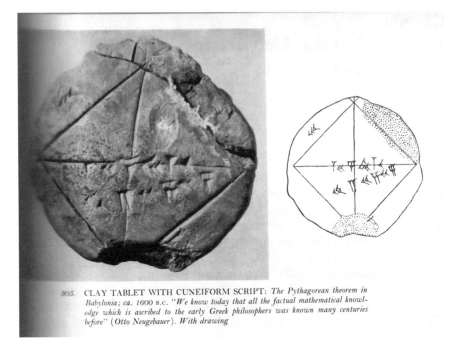

305. CLAY TABLET WITH CUNEIFORM SCRIPT: *The Pythagorean theorem in Babylonia; ca. 1600 b.c. "We know today that all the factual mathematical knowledge which is ascribed to the early Greek philosophers was known many centuries before" (Otto Neugebauer). With drawing*

Fig. 3.1 Clay tablet from Babylonia, 1600 b.c. from Giedion (1962) (Sigfried Giedion)

form of a *group* G, the group of rigid displacements. A "point" x in the space around us corresponds to the special sub-group of movements H_x that left x invariant. Then there is no real need of points, only the subsets of movements turning around them have a meaning in this approach. Consequently, the geometry comes from movements in time, ambiguous with respect to the division of the world between an interior and an exterior, and the space comes from the collection of displacements that don't escape to infinity.

Where this geometry is happening in the brain? Several signatures of Euclidean geometry are present in the primary motor area, other ones in the Basal Ganglia, or in the vestibular nuclei, and many of them at the «end» of the visual flow, in the para-hippocampal region, where are found place cells, head direction cells, grid cells, and so on. All the elements of classical geometry of Euclid are present in this region, but it contains also boundary cells, corner cells, velocity cells, time cells, thus many elements of dynamics and topology, as elaborated in particular by Poincaré in a mathematical theory.

In fact there is no unique definition of «geometry» in Mathematics, as it is the case for the term «energy» in Physics, according to Heisenberg (1969). However several successive levels are recognized today:

(1)   (from Galois and Klein) at a first level, a geometry is made by a collection of *sub-groups* H_x in a *group* G (Example: the group of displacements of a plane

Iso_ + (2) and the subgroups of rotations around the points x); the points in a
Klein space are defined by the transformations in G that fix them, as above;

(2) (from Riemann and Cartan) the preceding structure is retained only infinitesi-
mally; and the *curvature* appears as the obstruction to globalize G;

(3) (from Hamilton, Lagrange, Poincaré, Thom, Smale et al.) a *dynamic* or another
field is added (ex. a symplectic structure for classical mechanics);

(4) (from Grothendieck, Giraud, Verdier) sets are replaced by *topos* (ex: all maps
between sets according to a certain graph), then a constellation of geometries
can appear, operating coherently on a constellation of spaces. Elements of a
topos can be considered as *fields* over a background space, named a site.

Several examples of the first level can be seen in the brains of mammals.

For instance, as mentioned above, for the Euclidian geometry, O'Keefe and
Dostrovsky in 1971, discovered *place cells* in the hippocampus of rats, firing even
in the dark, when the animal is at a certain place in the plane, cf. Fig. 3.2.

Thus the rigid displacements are realized in the brain by the successive activations
of place cells (for coding translations) and by the pairs of *head direction cells* (for
coding rotations), discovered by James B. Ranck in 1984 (Fig. 3.2).

Hafting, Moser and Moser in 2005 discovered *grid cells* in the entorhinal cortex
near the hippocampus, which fire when the rat goes through any one of the vertices
of a regular lattice in the plane, cf. (Hafting et al. 2005). A group of cells share the
same translational invariance, with different phases.

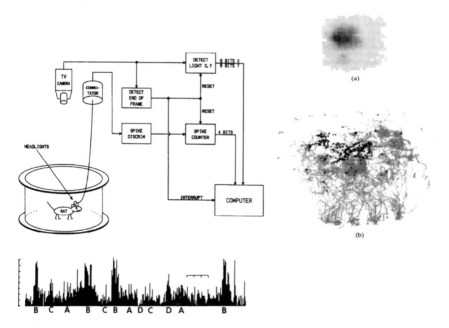

Fig. 3.2 From O'Keefe (1976)

The symmetry group Γ of grid cells fields is a discrete subgroup of the group of rigid transformations: (Fig. 3.3)

Geometry can predict the form of a 3D mesh for possible 3D grids (as it seems that bats possess). It is described in Fig. 3.4.

An example of level two, with curvature:

Non-linear deformation of the grids geometry can be seen as an occurrence of curvature effects (Fig. 3.5).

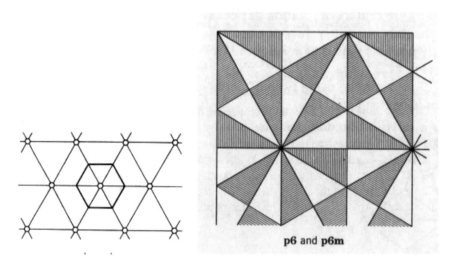

p6 and p6m

Fig. 3.3 From the book of Coxeter (1969)

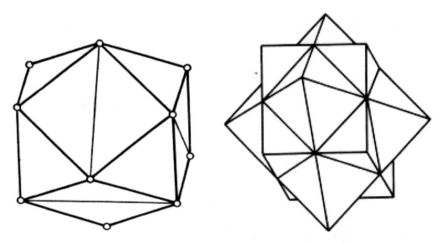

Fig. 3.4 Left (from Coxeter (1969)), the right explains how it is constructed from a cube and an octahedron

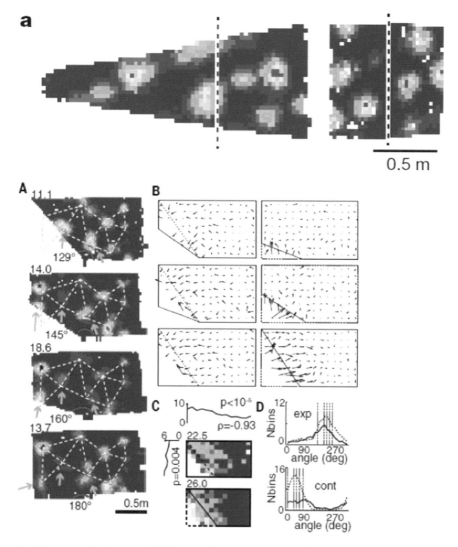

Fig. 3.5 a from Krupic et al. (2015) and **A, B** from Krupic et al. (2018)

As shown in Barry et al. (2007), the grid cells can adapt linearly their functioning to the form of the environment after stretching (but they do not adapt to a scale's change, as shown in Barry et al. (2012), cf. also (Hafting et al. 2005) (Fig. 3.6).

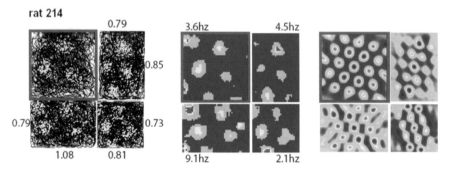

Fig. 3.6 From Barry et al. (2007)

3.2 Euler and Lie

Another example of level one is given by the implementation of *affine geometry* and its use in several brain's subsystems.

With respect to Euclidian geometry the affine geometry (recognized by Euler) forgets the notions of length and angle; it retains only the notions of point, line and plane, which can be secant or parallel.

The oriental art of affine is represented in Fig. 3.7 (illustration of Tosa Mitsuyoshi for the Genji monogatari, Japan, XI th century).

Fig. 3.7 Attributed to Tosa Mitsuoki (1617–1691)

When the environment of a rat is stretched, as we just saw, the grid cells adapt their firing as if the ambient geometry were Euclidian for a different measurement of lengths and angles. This is an affine adaptation, compensation from inside of an apparent change in the real world.

In the brain, the affine geometry contributes to organize the generation of voluntary movements, for writing and locomotion, in particular the relation between shape and timing: Handzel and Flash, 1997 (Handzel and Flash 1997, 1999) understood that the *two-third law* expresses the equi-affine invariance, a unique subgroup allowing computation. With Tamar Flash, Alain Berthoz and Ronit Fuchs in 2009, then with Cuong Pham, and more recently Yaron Meirovitch, we extended this analysis to the full affine group, incorporating a form a local isochrony.

Where in the brain? Higher visual areas like OT, premotor areas, cerebellum? nucleus NST? Note a deficit of full affine in Parkinson patients (Eran Dayan, Rivka Inzelberg and Tamar Flash, 2012, cf. Dayan et al. 2012) (Figs. 3.8 and 3.9).

With Tabareau et al. (2007), we showed that the complex conformal group contributes to organize the command of saccades in the Superior Colliculus.

Cerebral maps of the visual field in vertebrates show how different kinds of geometries, affine or conformal, can be used for organizing the functioning of eyes movements and more generally the reorientation of grasping or locomotion in space. The maps which represent the visual plane in the tectum of vertebrates, or the colliculus of mammals, can be linear (as for rodents) or logarithmic (as for cats, primates).

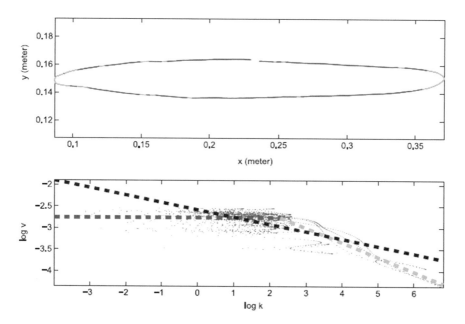

Fig. 3.8 From Bennequin et al. (2009) To account for the change in movement's timing along a long ellipse, it is better to assume a shift of Euclidian geometry to affine geometry than to look for an exponent law, $\log V = -\,b \log R + C$

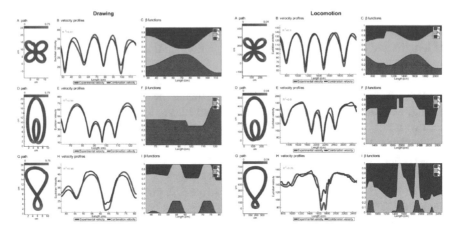

Fig. 3.9 From Bennequin et al. (2009) Colors show the respective contributions of Euclidian geometry (blue), equi-affine geometry (green) and full affine geometry (red), for drawing (left part of the figure) and locomotion (right part of the figure), for different shapes of trajectories. This shows the interplay between spatial and temporal aspects of voluntary motions

For certain animals like crocodile, the curvature appears, using approximate affine geometry for the horizontal motions of the eyes and head and more conformal geometry for their vertical motions, which reflects their life at the water surface of rivers (Fig. 3.10).

Note that the primary space of colors in the thalamus is also organized by an affine geometry in a 3D space, in particular for adaptation to illuminants; this geometry is suggested to help for color constancy, cf. (Bennequin 2014).

3.3   Kepler and Grothendieck

With D. Rudrauf, I. Granic, G. Landini, K. Friston, and K. Williford, in 2017 (cf. Rudrauf et al. 2017; Williford et al. 2018), we proposed that consciousness results from a shift from the 3D Euclidian and affine geometry to the 3D projective geometry, possessing points at infinity.The adaptation of projective frames can explain many illusions, for instance the Moon illusion (cf. Rudrauf et al. 2018) (Fig. 3.11).

Projective geometry can explain the very ancient observation that the moon appears much bigger at the horizon than higher in the sky. The reason is a transformation of the plane at infinity which results from a change in projective frame (five points in 3D space) induced by the maximization of visual information (Fig. 3.12).

In fact, this action of the projective group on the structures of information is directly related to the last level of geometry (up today) we mentioned above. Because

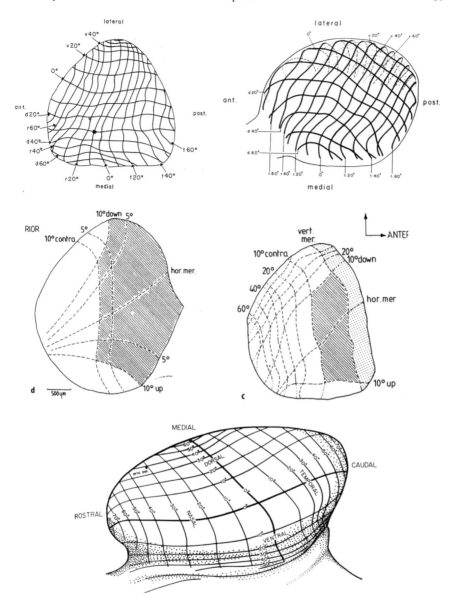

Fig. 3.10 Several retinotopic maps in the optic tectum of vertebrates or in the superior colliculus of mammals; from Kruger (1970) a rat and a fish, both being affine, from Cynader and Berman (1972), Feldon and Kruger (1970) respectively, a monkey and a cat, both being conformal, and from Heric and Kruger (1965) the optic tectum of an alligator, mixture of affine and conformal

Fig. 3.11 Ptolemy, from the book of Ross and Plug, Oxford 2002 (Ross and Plug 2002)

the PCM model of Rudrauf et al. is based on the principle of Free Energy Minimization (cf. Friston 2010), which involves a Bayesian network of observation and decision centers, giving a topos of information (cf. Baudot and Bennequin 2015; Vigneaux 2019), and asks that perceptions and actions evolve according to the minimum of a function of internal beliefs and goals, which is a trade-off between the conservation of a priori knowledge and the integration of novel data. The mathematical formula of this function is the sum of the expectation of energy and of minus the entropy (i.e.

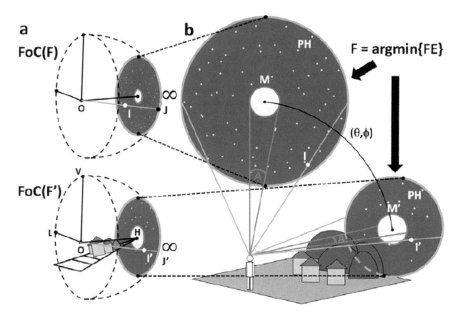

Fig. 3.12 From our paper on arXiv (Rudrauf et al. 2018), with its original legend Projective setup and argument

the negentropy) of the probability law which expresses the internal state of beliefs and goals.

3.4 Topos Geometry for Movements

More generally, with Alain Berthoz (Bennequin and Berthoz 2017), we have proposed that for the preparation and execution of movements, new kinds of geometries are necessary, without points, made by *topos* in the sense of Grothendieck and Verdier (cf. SGA 1972). Heuristically, a Grothendieck topos T is associated to a *site* S, made by nodes and by arrows between them, with a notion of coverings or refinements of the nodes by some families of arrows; an object X in the topos T is a functor over the site, i.e. the choice of a set X_s for each node s, and of a map between them f: $X_s \rightarrow X_t$, for each arrow t $\rightarrow$ s, working in a coherent manner. A topos is itself a category, where each arrow is a natural transformation between two objects, i.e. a collection of maps between sets F: $X_s \rightarrow X'_s$, compatible with the maps f from X_s to X_t and f' from X'_s to X'_t. In our case, to obtain a topos geometry, we choose in addition at each node of the site, a group, or better a *groupoid* (i.e. a category with all arrows invertible, but with several units), we link them by morphisms of groupoids, and we consider special objects of the topos, where the

groupoids act coherently, as being the geometrical objects. This gives a generalization of usual sets theory and of ordinary geometries, which is more flexible, local and
contextual. Concretely, the interesting sites for us are functional networks of body's
parts, articulations, neuromuscular junctions, ensembles of neurons and brain areas,
which organize, plan and control a special class of movements in the external space.

The five main examples that we consider are

(1) the own body space, or corporal scheme,
(2) the near space for grasping or reaching, or peri-personal space
(3) the near locomotion space, or extra-personal space,
(4) the space of navigation, or far environmental space,
(5) the imagined space, where our avatar can move.

For all these cases, we suggest that there exist a set A of interconnected areas,
executing or modulating the movements, and for each a in A, a virtual geometry, made
by a groupoid of transformations G_a, coherently acting on variable states X_a (for
instance quotient spaces G_a/H_a), which guide the adaptation of this area, in such
a manner that some of them command the specific actions in the physical space, for
instance in Euclidian or Galilean geometry. Entries in the network have two kinds
of sources, sensory variables and internal decisions. This network of geometrical
spaces forms a *topos geometry* covering G. This geometry has to be considered as
a unique coherent dynamical geometry, even if it manifests itself in each individual
area by different sets geometries.

In the simplest idealization, the underlying *site* A can contain just an arrow (plus
two identities), but more realistically A possesses many vertices and loops, working
together for perception/action. Remark: for describing correctly the dynamic in the
topos it is necessary to include time, the site is note static.

Take for instance locomotion along a complex path, with obstacle avoidance.
The superior colliculus and related subcortical and cortical areas is working in
affine/conformal geometry for anticipatory gaze movements and head or body direction changes; it is a part of the support of visual adjustments and body orientation,
another related source for that is the parietal cortex, which is perhaps more connected
to projective geometry, manipulating frames attached to external cues or obstacles
in addition to the self. The vestibular system is working in a Galilean geometry,
acting on the space of angular velocities and linear accelerations; it contributes to the
postural and equilibrium control, sending information to the spinal neurons, directly
or in a loop involving the cerebellum. The principal loop in the brain for selecting
the behavior, in locomotion as for the other voluntary motions, involves the basal
ganglia (BG), the Thalamus and the motor cortex; as revealed by cell's receptive
fields, this loop is probably working mostly in Euclidian geometry, but it is also able
to integrate affine elements, because it determines total duration and timing modulation of the movement, that we know influenced by these non-euclidian geometries.
With Tamar Flash and Alain Berthoz, taking in account perception experiments (cf.
Dayan et al. 2007), we suggest that the dorsal premotor areas in particular can anticipate the movement by working in equi-affine geometry. The motor cortex and BG
(and deeper regions like the Hypothalamus) project to the centers of commands in

the midbrain, MLR (mesencephalic locomotor region) and DLR (diencephalic loco-motor region), which themselves project to the reticular formation (RF) in the brain stem, which finally project to the spinal cord, where are the motoneurons. Through other brain stem nuclei, the cerebellum intervenes for generating precise movements and corresponding adapted postures of all the body's segments; we suggest that the cerebellum is working also in a geometry which is wider than the Euclidian one, like the affine or equi-affine geometry (cf. Habas et al. 2020). However, we can expect that totally new geometries appear in the network, when approaching the execution centers, for instance MLR, DLR, RF, Spinal cord. These geometries will take care of the mechanisms of muscular commands, forces and torques, they will not necessarily be represented by groups, they could use *groupoids*, expressing the non-holonomic constraints of real body motions. These new geometries will define postural spaces, all that being necessary for dimensional reduction and adaptation.

In fact, we can imagine two degrees of freedom for the topos geometries, in one direction we vary the network A, in the other one we vary the groupoids and sub-groupoids constituting the geometry at each area in A. When convenient gluing conditions are verified, the set of groupoids over A forms a *stack* (champs in French) in the sense of Grothendieck and Giraud (Giraud 1971). This opens to the possible introduction of curvature in this setting.

Consider several topos associated to different tasks; they must be linked together by specific mappings, named functors, which are themselves linked by natural trans-formations between functors. This is what happens theoretically for the class of all Grothendieck topos, forming a structure named a 2-category. In the context or action/perception, this corresponds to the necessity of combining different sorts of actions, like walking for grasping something and at the same time planning how to find his way in the forest.

Alain Berthoz had suggested before that several geometries are necessary for guiding several networks controlling actions in different spaces; cf. (Berthoz 2003, 2011, 2016). With the 2-category of topos, it appears a set of different but compatible geometries that cover respectively the body's movements, the motions of the objects, the trajectories of the center of mass in medium/near space or along a curved path of twenty meters, or the route/fly to a far imagined mountain.

We are working now with Alain Berthoz to make more precise these new topos geometries. For that we have collaborators in several domains, in particular Tamar Flash for geometry and motor control, Jean-Paul Laumond for robotics, the group of Rennes, Anne-Hélène Olivier, Armel Cretual, Julien Petre, on locomotion in near and far spaces, and François Garnier, on visual arts, in particular movies and virtual reality (Fig. 3.13).

Imagine connected geometries on each vertex of this graph.

Now, consider the first brain along the line going to vertebrates: the tunicate larva, with one eye, one otolith, a chord and muscles. It has a modular brain, with several sensory and motor subsystems. Thus the origin of the brain is for controlling movements, and perceiving space (as said by Rodolfo Llinas, in his book Llinas 2001) (Fig. 3.14).

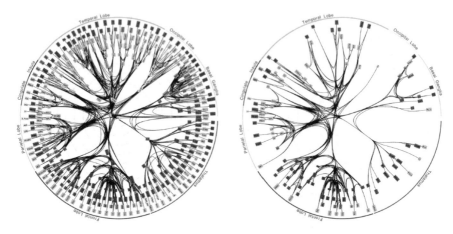

Fig. 3.13 The innermost core of the primate brain, according to Modha and Singh (2010)

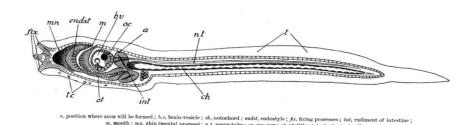

a, position where anus will be formed ; *b.v*, brain-vesicle ; *ch*, notochord ; *endst*, endostyle ; *fix*, fixing processes ; *int*, rudiment of intestine ; *m*, mouth ; *mn*, chin (mental process) ; *n.t*, nerve-tube ; *oc*, eye-cup ; *ot*, otolith ; *t*, test ; *t.c*, test cells.

Fig. 3.14 Free swimming tadpole of the simple Ascidian *Ciona intestinalis seen* from the side (combined from two figures by Kowalevsky)

However the brain has evolved for emulating more and more activities, not only movements planning, but imagined complex social relations, explicit or implicit reasoning and imagination, concepts and arts. In all these domains, as for planning and modulating motions, virtual spaces are created inside the brain, equipped with ideal geometries for guiding adaptation. Following works of Grothendieck, Brieskorn, Thom and Looijenga, we have suggested that dynamic functions, control parameters, and (homological) geometry for adaptation, are organized by dynamical *ternary structures*: (1) a dynamical system X, realizing the function (information transmission, nerve impulses, motor commands, ...), (2) a manifold Λ, named the unfolding of X, representing the parameters that modulate X; (3) another space H, named the homology of X over Λ (more precisely its vanishing co-homology), which is implemented by real cells, but has a virtual function, a sort of observer, linking X and Λ (cf. Slotine and Lohmiller 2001). The inner geometry is on H, and controls the forms of deformations in Λ. The space H understands what happens as essential in X, and controls the paths of deformations in Λ, for adaptation.

All these spaces exist as well for topos, replacing sets by fields (or by stacks).

Fig. 3.15 From a book (Bompard-Porte and Bennequin 1997) written with Michèle Bompard-Porte, describes the ternary dynamics as a wheel

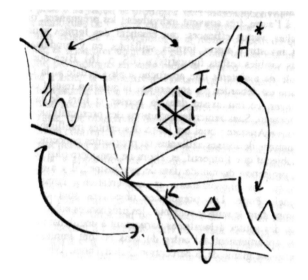

Fig. 3.16 Notre Dame of Paris. This is a rare picture (sculpture) which allows us to feel directly the movements of bodies in space

In humans, imagination and symbols reproduce such structures (Fig. 3.15).

Every work of art is virtually a transformation of us, which makes active our body and our imagination (Fig. 3.16).

Acknowledgements I thank warmly Tamar Flash and Alain Berthoz for their invitation to this wonderful meeting on Arts and Movement Science. The following notes and images closely follow the oral exposition.

References

Barry, C., Hayman, R., Burgess, N., & Jeffery, K. J. (2007). Experience-dependent rescaling of entorhinal grids. *Nature Neuroscience, 10* (6).

Barry, C., Ginsberg, L. L., O'Keefe, J., & Burgess, N. (2012). Grid cell firing patterns signal environmental change by expansion. *PNAS, 109*(43).

Baudot, P., & Bennequin, D. (2015). The homological nature of entropy. *Entropy, 17*(5).

Bennequin, D. (2014). Remarks on invariance in the primary visual systems of vertebrates. In G. Citti, A. Sarti eds, *Neuromathematics of vision*. Berlin: Springer.

Bennequin, D., & Berthoz, A. (2017). Several geometries for movements generations, in geometrical and numerical foundations of movements. In Laumond, J.-P., Mansard, N., & Lasserre, J.-B. (Eds.), *Springer tract in advanced robotics* (Vol. 117, pp. 13–43).

Bennequin, D., Fuchs, R., Berthoz, A., & Flash, T. (2009). Movement timing and invariance arise from several geometries. *PLoS Computational Biology, 5*(7).

Berthoz, A. (2003). *La décision*. Paris: Odile Jacob.

Berthoz, A. (2011). *Simplexity. How to deal with a complex world*. Yale University Press.

Berthoz, A. (2016). *La vicariance*, Odile Jacob 2013, translation *Vicariousness*, Harvard University Press.

Bompard-Porte, M., & Bennequin, D. (1997). *Pulsions et Politique, une relecture de l'évènement psychique collectif à partir de l'œuvre de Freud, suivi de Le non-être homologique*. Espaces Théoriques: L'Harmattan.

Coxeter. H. S. M. (1969). *Introduction to geometry*. Wiley.

Cynader, M., & Berman, N. (1972). Receptive-field organization of monkey superior colliculus. *Journal of Neurophysiology, 35*(2).

Dayan, E., Casile, A., Levit-Binnun, N., Giese, M. A., Hendler, T., & Flash, T. (2007). Neural representations of kinematic laws of motion: Evidence for action-perception coupling. *PNAS 104*(51).

Dayan, E., Inzelberg, R., & Flash, T. (2012) Altered perceptual sensitivity to kinematic invariants in Parkinson's disease. *PLoSone, 7*(2).

Feldon, P., & Kruger, L. (1970). Topography of the retinal projection upon the superior colliculus of the cat. *Vision Research, 10*(2).

Friston, K. (2010). The free-energy principle: A unified brain theory? *Nature Reviews Neuroscience, 11*, 127–138.

Giedion, S. (1962). *The eternal present, a contribution on constancy and change*. Oxford: Oxford University Press.

Giraud, J. (1971). *Cohomologie non abélienne*. Berlin: Springer.

Habas, C., Berthoz, A., Flash, T., & Bennequin, D. (2020). Does the cerebellum implement of select geometries? A speculative note. *The Cerebellum, 19*, 336–342.

Hafting, T., Fyhn, M., Molden, S., Moser, M. B., & Moser, E. I. (2005). Microstructure of a spatial map in th entorhinal cortex. *Nature, 436*(7052).

Handzel, A. A., & Flash, T. (1997). The coordinates of the binocular motor system. *Abstracts of the Society for Neuroscience*.

Handzel, A. A., & Flash, T. (1999). Geometric methods in the study of human motor control. *Cognitive Studies, 6*(1–13).

Heisenberg, W. (1969). *Der Teil und das Ganze*. Munchen: Riper and Co.

Heric, T. M., & Kruger, L. (1965). Organization of the visual projection upon the optic tectum of a reptile (alligator mississipiensis). *Journal of Comparative, 124*, 101–111.

Kruger, L. (1970). The topography of the visual projection to the mesencephalon: A comparative survey. *Brain, Behavior and Evolution, 3*, 169–177.

Krupic, J., Bauza, M., Burton, S., Barry, C., & O'Keefe, J. (2015). Grid cell symmetry is shaped by environmental geometry. *Nature, 518*(232).

Krupic, J., Bauza, M., Burton, S., & O'Keefe, J. (2018). Local transformation of the hippocampal cognitive map. *Science, 359*(1143–1146).

Llinas, R. (2001). *I of the vortex. From neurons to self*. MIT.

Modha, D. S., & Singh, R. (2010). Network architecture of the macaque brain. *PNAS, 107*(30).

O'Keefe, J. (1976). Place units in the hippocampus of the freely moving rat. *Experimental Neurology, 51*(1).

Poincaré, H. (1902). *La Science et l'Hypothèse*. Paris: Flammarion.

Ross, H., & Plug, C. (2002). *The mystery of the moon illusion*. Oxford.

Rudrauf, D., Bennequin, D., Granic, I. Landini, G., Friston, K., & Williford, K. (2017). A mathematical model of embodied consciousness. *Journal of Theoretical Biology, 428*.

Rudrauf, D., Bennequin, D., & Williford, K. (2018). The moon illusion explained by the projective consciousness model. *Journal of Theoretical Biology, 507*(2020), 110455.

SGA IV. (1972). *séminaire du Bois Marie, 1962–64*. Berlin: Springer.

Slotine, J.-J. E., & Lohmiller, W. (2001). Modularity, evolution and the binding problem: a view from stability theory. *Neural Networks, 14*, 137–145.

Tabareau, N., Bennequin, D., Slotine, J.-J., Girard, B., & Berthoz, A. (2007). Geometry of the superior colliculus mapping and efficient oculomotor computation. *Biological Cybernetics, 97*(4).

Tosa, M. (1617–1691) Illustration for the Genji monogatari, Japan, XVIIth century.

Vigneaux, J.-P. (2019). Topology of statistical systems. A cohomological approach to information theory, Thesis, Paris VII.

Williford, K., Bennequin, D., Friston, K., & Rudrauf, D. (2018). The projective consciousness model and the phenomenal selfhood. *Frontiers in Psychology, 9*.

Chapter 4
Traces of Life

Thierry Pozzo and Juliette Pozzo

Abstract   The goal of this paper is to investigate the relationship between Art and Science in the light of EJ Marey's work and the invention of chronophotography. We begin historically with Marey's technical inventions, allowing for a representation of biological changes by freezing humans, while walking, while horses are galloping, or while birds are still in flight. From these traces of life Marey built prototypes made up of generic behaviors, and created a new visual language for artistic and scientific communities. We hypothesize that the lines separating Art and Science tend to disappear as soon as they share the same pictorial medium. We then go on with recent empirical approaches describing the visual process at work during the perception of human movement, an experiment based on Marey's methodological innovation. We show that visual displays of a shapeless walker brought significant sensorimotor information, a result confirming that the motor system plays a crucial role in visual perception, especially when only a few dots are visible. Finally, we turn to a crucial question for both the scientist and the artist: how can an abstract and impersonal trace rendering biological movement reach the field of the observer's private representations? We conclude with the notion of singularity in Science and Art.

Electronic supplementary material   The online version of this chapter (https://doi.org/10.1007/978-3-030-57227-3_4) contains supplementary material, which is available to authorized users.

T. Pozzo (✉)
IIT@UniFe, Center for Translational Neurophysiology, Istituto Italiano Di Tecnologia, Ferrara, Italy
e-mail: thierry.pozzo@iit.it

INSERM U1093 Cognition-Action-Plasticité Sensorimotrice, Faculté Des Sciences Du Sport, Campus Universitaire, BP 27877, 21078 Dijon, France

J. Pozzo
Musée National Picasso-Paris, 20, rue de la Perle, 75003 Paris, France

T. Flash and A. Berthoz (eds.), *Space-Time Geometries for Motion and Perception in the Brain and the Arts*, Lecture Notes in Morphogenesis, https://doi.org/10.1007/978-3-030-57227-3_4

4.1 Introduction

If we go far back into the past, and look around the time of the renaissance, we find little difference between the point of view of the artist and that of the scientist. An occidental academic apartheid then isolated these two domains that today seem to be trying to recombine, making it more difficult to view Art and Science as independent specialities. The important use of technology for contemporary artistic creation has encouraged new collaborations and contributed to the emergence of hybrid cultures, where artistic fantasy mixes with technical skills: Art and Science being now immersed in a same cultural agenda, rooted in techno-science. Moreover, the abundant use of pictures and diagrams to communicate experimental results has introduced an aesthetic impetus to scientific production. Humanoid robotics, styling its results on the model of Hollywood science fiction creatures, exemplifies a 'spectacularisation' of science. The computer graphic artist who covers the cyborg with a soft human skin blurs the boundaries between a human, a robot and a digital avatar. Bio-inspired on the surface, the humanoid robot is a model of nature but still very different from nature itself. The overuse of images, as efficient "all-at-a-glance charts" contrasts with long columns of numbers (Tufte 1997), so bringing out complex activities which would otherwise lie unobserved, and which certainly allow for a greater scientific presence in the social space. But we still run the risk of promoting form at the expense of function, so attaching the spectator to the envelope rather than to its contents.

Bringing Art and Science closer does not prevent the persistence of taboos, prohibitions which make it difficult to consider how aesthetics has been introduced into science, or what role emotional processes play in scientific activity. Unlike the recent enthusiasm for neuro-aesthetic studies and the description of brain activities induced by aesthetic experience, the literary value of a scientific paper and the aesthetic component of its diagrams are rarely examined. Nevertheless, studying the way science works, as can be done through an examination of science fiction literature, which communicates knowledge according to a stylized mode of writing, might help to improve our understanding of the hidden but intense links between the two domains, as also to make it easier to distinguish a human from a cyborg, that is, to separate realistic knowledge from imaginary knowledge (Elias 2016).

This two-way contamination between Art and Science becomes clearer if one takes the trouble to go back to the origin of scientific graphics. The work of Etienne Jules Marey, one of the most important contributors to the development of scientific imagery, has generated a useful corpus of data to investigate these relationships. As we shall see, focusing on the seminal use of photography in science helps to answer recurrent questions about the interactions between Art and Science: How the use of an artistic media (photography) has precipitated the interactions between Art, inasmuch as it seeks to express subjectivity and Science, inasmuch as it seeks objectivity? How has the de-compartmentalization of these two specific human activities been promoted? Here, the invention of chronophotography proves to be decisive for our understanding of the way in which these two domains have been brought together.

Chronophotography and kinematic visualization have both given the scientific community the tools they needed to quantify visual impression hitherto subjectively written down in the physiologist's notebook. The success and diffusion of this new way of describing the animal kingdom nevertheless introduces a standardization of representations, bringing uniformity alongside the multitude of its past subjective depictions. A brief examination of Western visual arts since the end of the nineteenth century highlights such a normalization of visual perception. More important, the possibility of bringing the functioning of organs into the focus of immediate visual perception has done much to transform science into a kind of theater, leading Westerners to raise questions about, and to explore in a new way, both their relationship to reality and their artistic practices (Pozzo 2003). Scientific projects, and artistic creations based on a method that makes the hidden phases of biological movements more visible, are here considered as evidence supporting the view that scientific and artistic thinking (see Tversky 2015) are coming together, this at a time when the rationalization of human work is being accelerated by technical progress (Gibret and Gilbret 1917).

We begin historically with Marey's technical inventions, which began with a holistic approach, and so could not avoid a reductionist approach threatening to split human behavior into elementary units. We then deal with the diffusion of this pictorial revolution into the artistic community. We go on with recent empirical approaches describing the visual processes at work during the perception of human movement, a field of research that has become possible as a result of Marey's methodological innovation.

Finally, we turn to a more epistemological question concerning the interaction between Science and Art: How one uniquely standard trace, cleansed of any human shape and devoid of subjectivity, can induce emotion, that is, an individual human reaction relying on specific cultural and educational components? And how artistic subjectivity can itself be transformed into an object for science, a domain dedicated to revealing universal rules masking individual singularities?

4.2 Body Geometrization and Dissolution of the Living in White Dots

The impact of photography on human physiology cannot be fully understood without first considering Marey's preceding graphical method (Fig. 4.1), set out in a continuum including graph and chronograph as abstract and discrete temporal charts for chronophotograph as a photographic representation of motion over time (Marey 1884; Frizot 2001). Without traces of the observed phenomenon, the physiologist must listen, observe and then report, verbally or in writing, on subjective feelings. To deliver an objective memory of natural events, his initial postulate is very simple: any biological function results from a mechanical or chemical cause, which, thanks to a battery of devices, is translated into temporal and spatial signals. Following this

Fig. 4.1 A artistic representation from Marey's graphical method. Horse and Rider with inscribing apparatus: Rubber balls are attached under the horse's shoes (right part) and connected to a recording device by rubber tubing fastened along of its legs in order to portray the rhythm of the gaits of horse

principle, the vital functions are flattened like the painter's work on the canvas, and then converted into spatiotemporal variables, in the form of Cartesian coordinates. Frogs, insects, birds are directly connected to graphic machines that trace patterns of wings or legs onto the surface of smoke-blackened cylinders (Marey 1873).

Distinct from his pioneering graphical inscription demonstrating natural phenomenon under abstract graphical representation (Fig. 4.2), Marey's first motion portrays events in time recorded outdoors and showing recognizable actors (Marey 1892, 1899). By creating a stroboscopic-like effect, Marey, with the decisive help of his assistant Georges Demeny, transforms smooth biological movements into successive postures: the static replaces the dynamic, each cliché suspends time, and the actors are devitalized. This type of picture, based on an accumulation of body positions, had already been conceived as soon as humans were able to transform a mental state into a drawing. Indeed, the images from the Chauvet Cave in Southern France, thought to date back 30–32,000 years ago (Sadier et al. 2012), and regarded as the oldest abstract animal in motion figuration (Azema and Rivere 2012), clearly depicts overlapping body parts remarkably similar to an agglomeration of modern snapshots (See Chabrier for a fictive and artistic animation of these visual relics) and foreshadowing Rodolphe Töpffer (1833) and what transpires later in pioneering comics. The presence of pictures of animal in action in the prehistoric world testifies

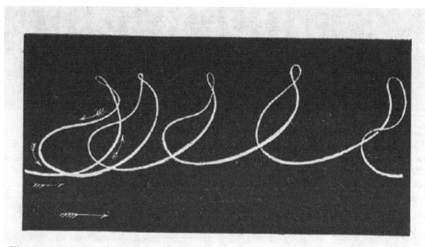

Fig. 34. — Trajectoire de l'extrémité de l'aile d'une corneille. Une paillette brillante attachée à la 2ᵉ rémige suivait le parcours indiqué par de petites flèches courbes. En bas de la figure une flèche droite et horizontale exprime la direction du vol.

Fig. 4.2   Trajectory of the end of a crow's wing, in EJ Marey, La photographie du movement (1892), G Carre Ed., Paris (pp 74)

to the biological roots, in humans, of the urgent need to translate an act into its visual expression. As many others before, EJ Marey also succumbed to just such an implicit reward, enabling him to prolong the pleasure of a past observation, the sort of reward which might have prepared humans to catch a prey, or avoid a predator, all over the course of their evolution.

Although he moves the lab into the open-air (see also E Oehmichen investigations), Marey's thought remained that of a convinced positivist relying on a rational empirical approach: start from the complex, go towards the simple, and give an intelligible explanation, first by geometrizing the body, then by converting geometrical abstractions into numerical values. In a universe of movement for which there was no adequate source of measurement, EJM first showed and quantified the complexity (from the human body to smoke filament analysis and aerodynamic phenomenon) by plotting the movements on a 2D sheet.

Chronophotographic charts and the outlines of humans in motion are exemplary visual testimonies of Western scientific thought, committed to rationality and the quest for human efficiency. It is no longer necessary to talk about nature, but rather to let it speak for itself with pictures that transform nature into a sincere testimony (see Fig. 4.3). The once subjective observer of a singular lived experience now becomes a spectator of a visible outer world, one which could be formalized mathematically. As such, chronophotography perfectly fits within a mechanistic approach. Accordingly, the laws deduced from the observation of the non-living are transposed over to the living body to describe and explain it (Chazal 2008). And this no matter what the scale

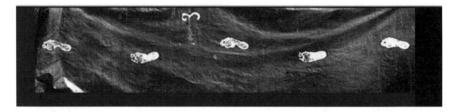

Fig. 4.3 Photography of locomotion traces as an objective memory of natural events: footprints on a black curtain from a walker moving forward from the right to the left or walking backward in the reverse direction. Note that the visual perception of such traces are strongly evocative of a walker in action when they are visible as a whole instead of isolated (Ref Cinémathèque Française, Cote de numérisation: P069-024) http://www.cineressources.net/ressource.php?collection=PHOTOGRAP HIES_NUMERISEES&pk=42470

of the phenomenon or its level of complexity, both macroscopic and microscopic, human or bacterial. The task is ambitious, since a physical description of the basic elements constituting biological organisms must account for their vital functions, as if the knowledge of the molecular structure of water could reveal the secrets of the hidden life of the oceans and its streams!

Compared with earlier graphical methods, chronophotography avoids the material chain linking the phenomenon to the recording device. An ancestor of modern telemetry, the electromagnetic signal of photography leaves a chemical trace inscribed onto the sensitive film without any intermediary. Free of potential sources of interference and noise induced by the units of a complex recording set-up, the motion-capture brings authentic footprints of gestures without hindering their execution. Resulting from a typical method strongly impregnated with the positivism of A. Comte (i.e., exploring and numbering a natural phenomenon disconnected from its milieu), the millisecond-by-millisecond portrayals of body movements frozen in time then provides remote and extrinsic evidence devoid of any human contact, evidence generated by traditional laboratory manipulations. As a kind of epistemic apotheosis of the experimental method, it is supposed to comprehend nature by decomposing it, using photography as "an artificial retina" (Hoffmann 2013), and placing it in front of the eyes. Two principles guide the scientist here: (a) to reduce the complexity of the observed phenomenon to computable and intelligible digits, and (b) to avoid the subjective verbal report of the observer, exiled in Cartesian territory (see Bitbol 2010) and transformed into a silent spectator, distant from the object and disengaged from the transcript of the phenomenon. From this, Marey created an *environment* ("to be separated from") and escaped from the *milieu* ("to be in symbiosis with").

Regarding Marey's reductionism, his empirical inquiries start with a series of snapshots full of exotic details revealing the geographical and social origin of the actor: a running athlete of the *bataillon de Joinville* dressed in a white suit embellished with a cap and girdled with a large black silk sash, or the gallop of an Arab rider draped with a bubbling *burnous* and wearing a hat with a wide turban (Fig. 4.4a), or a walker in *canotier* cramped into a pea jacket of raw cloth, spanning the clock that gave the passage of time to the experimenter (Fig. 4.4b). Then, EJ Marey gradually

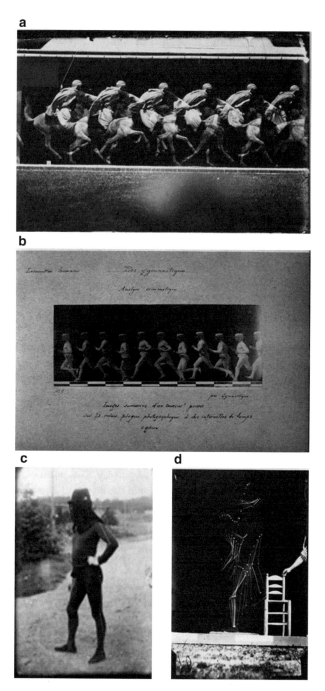

Fig. 4.4 A series of Marey's snapshots. From **a** to **d**: running athlete of the bataillon de Joinville; gallop of an Arab rider; a walker in canotier; subject covered with a black suit contrasting the white markers to be recorded; successive body postures recorded during a jump down with Marey's chronophotograph

undresses his models to make them universal: the actor easily identifiable by the dress codes vanishes below an undefined mannequin outfitted with a black body stocking (Fig. 4.4c), later underlined with white stripes joining the limbs. Body shape is then reduced to a stick figure of only one hemi-body (Fig. 4.4d), a skeletal figuration finally reduced to several white dots distinctly appearing against a black background. Notably, this cascade of reductions in 3 steps, going from a 3-dimensional body (a full actor) toward a 2-dimension skeletal configuration, and finally into a 1-dimension dot (that refers here to a "third order body metaphor", that is a bodiless image) nicely anticipated the so-called transition from analog to digital technology.

As a pre-requisite for any quantification, life has to be paralyzed into a photographic series and limited to few residual signs. The use of statistical tools will prolong the effort of disembodiment, transforming the Arab rider, or the walker in *canotier,* into an impersonal model from whom it is by now possible to extract, and average out, kinematic regularities. The power of a scientific law being proportional to its ability to generalize to everyone, the style and origin of the rider disappear in favor of an archetypal illustration of the law. Interestingly, such a dissolution of singularities into the form of an average human has certain similarities to that of Galton's composite portraits. At about the same period, Galton's method merges a multiplicity of individual clichés (the ancestor of digital morphing), into one single picture, in order to isolate a physiognomic model. The robot portrait to which members of a family line are related, the prototypical face of criminals or syphilitics are reconstructed under the superimposition of the photographic negatives, fixing each element reflecting the group of people. This is reminiscent of modern brain imaging method using aggregate statistics, and which looks for the mean group effect, defined as the ultimate standard individual where intersubjective variability is considered as noise (Seghier and Price 2018).

About the second principle and the claim for more objectivity, the discrete imprint of the gesture now concentrates the real into a digit. The transformation of a social character to a skeletal body configuration, and lastly to a number, is nonetheless achieved at the expense of a huge compression of the original. The method described as "radical empiricism", following William James parlance, distances the observer from the actor: the study of *a walking man* becomes that of *the human walk* (Didi-Hubermann and Mannoni 2004). Marey catches natural phenomenon from a third person perspective, which denies itself or tends to be forgotten, and where the overlooking interpreter is absent from the natural milieu (Bitbol 2010). But does the image sincerity claimed by a convinced positivist suffice to get us to believe that the photographic trace guarantees objectivity? It is most likely that Marey's analysis has been biased by the aesthetic emotion of the experimenter discovering a previously invisible world, and fascinated by the capacity of his photographic machine. Finally, the scientific photographer records and shows a domesticated nature fixed from one particular point of view, that of the one who is behind the camera, who frames the scene, decides the lighting, chooses and stages the actors, all of which figure as so many intentions rich in personal choices (Pozzo 2013). See Fig. 4.5.

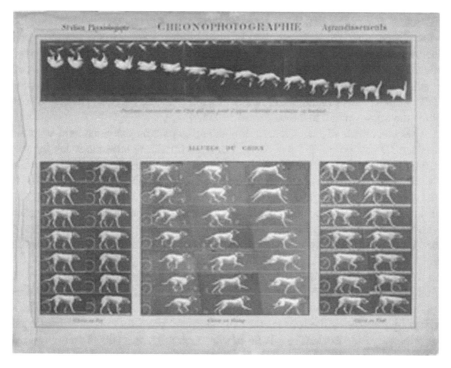

Fig. 4.5 Example of chronophotographies illustrating Marey's atlases. Upper row, falling cat; lower part, dog's locomotion

4.3 The Calibration of the Occidental Eye

As soon as scientists found a way to keep track of past motions, a quantitative physiology was possible. This opportunity was first adopted by an English photographer EJ Muybridge, and then by EJ Marey (1830–1904). Muybridge, who was Marey's exact contemporary (they were born and died within weeks of each other in addition to sharing the same initial), first disclosed the true nature of each horse's particular gait (walk, trot and gallop). However, the battery of classical cameras he used was incompatible with a reliable normalization in time and space, and with precise measurements. Most often, his concern was to build narrative representations instead of motion analysis, where models are performers playing inside schematic decors. Each picture was thus re arranged in an order following the performance asked of the actors. When a photo was missing, the previous, or the following, snapshot replaced the missing one, leading Muybridge to make the first photomontage in the history of the cinema (see Braun 1992). In contrast to Marey's motion capture, Muybridge's juxtaposition of pictures is not a collection of body postures imposed upon one single support, but separate images taken with different cameras and points of view. The short story reconstructed by laying each successive cliché end-to-end, inevitably disrupts motion fluidity, and so is perceived as discrete pieces, rather like what virtual photographers, standing in front of each camera in the battery, would themselves perceive looking into the viewfinder of their own camera. While Marey's stick diagrams convey generalities, Muybridge's photo-romance tell us stories with isolated pictures. Remarkably, Muybridge and Marey's opposite professional careers illustrate how science and art are interacting but not merging. One wanted to show while the other wanted to know (Pozzo 1995a, b).

These technical advances had no immediate impact on the behavioral sciences. Modern 3D motion capture devices will only be fitted into physiology laboratories at the end of the twentieth century, but the worm was in the fruit, especially in the artistic field. As soon as movement could be caught in an instant, the description of living beings took a giant step. Indeed, before motion capture devices were able to fix successive body postures, the representation of human locomotion disagreed with modern realistic representations. Like the first galop by Meissonier (see below), the Weber brothers' drawings (around 1850) exemplify the subjective experiential approach of biomechanics before the use of chronophotography. Indeed, the loco-motive pattern is inadequately depicted with exaggerated knee flexion and the trunk tilting forward (Mannoni 1999). Remarkably, the inadequate body posture sketched by the physiologists corresponds to the stance phase, where the reaction force trans-mitted by the sole of the foot is at a maximum and provides important tactile and proprioceptive feedback. The advent of chronophotography allows an appropriate time for reflection. The living world, petrified in the form of an immobile collection of stances, is no longer described on the basis of a mental rehearsal of kinesthetic sensa-tions. Action portrayal, poorly attached to the body of the observer, is now directly given by means of a machine interposed between the observer and the ecosphere. In some way, Marey's pictures are supposed to cool down the physiologist's emotion, an intention nicely worded by Cutting (2002) as follows: "*before the advent of the photographic instant there was the artistic moment*".

The diffusion of abstract kinematic traces standardized a way of looking at the living world, probably just as much as the invention of the Telegraph drastically re-scaled subjectively the distances and our representation of communicative space. Biological motion becomes visible beyond what the eyes can sample, allowing for a representation of living permanence that no longer has to be imagined. Further, after chronophotography has managed to replay life at the stroboscope's frequency, human visual sensitivity to repetition went from astonishment to indifference through gaze blunting and the recurrent observer's confrontation with repeated spectacles of repetition (Debord 2002). Just as Picasso taught us to recognize women's bodies in sketches for which most of the contours are lacking, EJ Marey taught us much earlier to see abstract figuration, which however teaches us nothing about ourselves but simply shapes our perception, and then transform us.

4.4 When Science and Art Breathe the Same Atmosphere

Chronophotography normalized the point of view from which all things were seen, and so created a real visual revolution by putting a grid in front of our eyes as we looked at animals and humans. A real visual revolution, like the one that normalized the point of view, with chronophotography putting a grid in front of our eyes as we looked at animals and humans. This created a public space of knowledge, a community of beliefs and shared conviction, making possible measurement, classification and comparison, all organized in accordance with the same language, based on one and the same body of data. Marey and Muybridge's atlases (Fig. 4.4) displaying canonical views of bodies in motion, printed and disseminated all over the world, are read by people from different cultures, who now see the animal kingdom differently. The serial overlapping of instantaneous postures of a running horse or a walking man becomes a formidable database for the artist, at the risk however of confusing Art and Science, and so reducing the canvas to pale imitations of decomposed and far too realistic movements. Auguste Rodin and Charles Baudelaire both asked: do masterpieces have to represent the real? Indeed, by displaying mechanical reality, photography informs the artist but makes of the masterpiece a *nature morte* mainly reproducing its visible shape.

A first wave of contamination reaches realistic painters. For Ernest Meissonier (1815–1891), artwork should go beyond the naive vision of the painter who depicts explosions, energy and the power of the animal. The contagion was so strong that after learning of Muybridge's horse-gallop decomposition, Meissonier decides to resume his artwork and correct it. Notably, the corrections made by the painter to his first version of his galloping horse will refer naively to a *"repentir"*. For the picture that shows the biomechanical truth transforms the obscurity of sensations into a new intelligibility. Thanks to motions capable of capturing the gallop of horse's hooves as a universal prototype, the naturalistic school would now be able to deliver true sensory signals instead of creating a subjective feeling.

A second wave reaches the "avant-garde" of the early twentieth century. One the most famous artworks inspired by the photographic display of movement over time is the "nue descendant l'escalier" by M. Duchamp, a mischievously artistic experimenter. He was a kind of "subjective scientist" as well as being an "objective artist", who chose derision in much the same way as Dada, to torpedo the myth of progress, which was supposed to bring happiness to all. In the same spirit Man Ray, strongly inspired by Marey's body geometrization, creates artworks for a more dreamlike and humorous purpose (e.g., "L'homme d'affaire"). At the opposite end of the spectrum, we find the Italian futurist school and so-called "photodynamism", where the chronophotograph provides a key to depicting changes and to conveying the illusion of movement in addition to the "vibration of the modern life" (Bragaglia 1913). The features of human facial expression no longer matter because the motion must now dominate the whole character. This artistic option fully coincided with a trend encouraging the development of science and technology for the production of art.

Among many other examples of artworks that echo the work of Marey and Muybridge, we should also mention Georges Seurat, who incorporated the ornamental, repetitive rhythms and the flattened space of the chronophotography in his "Chahut". Similarly, though he is more interested in creating an artificial world than using the film to reproduce natural facts, Edgar Degas took up again the idea of reappearance in his work. As did also Picasso and the Cubists, who specifically included features such as the immediate switching between multiple perspectives (a kind of cubism in motion) that contrasted with Marey's series depicted inside a fixed frame of reference. More recently, the contagion has been continued with Pop Art, or the art of painting based on those luminous traces that moving objects leave behind them (see Gjon Mili's photography).

Although obvious here, the influence of science on art is not unambiguous. At the beginning of the twentieth century, artists and scientists experience an intoxicating fascination for the machine, each in his or her own way. For the former, the body takes on the appearance of a machine, for which (s)he feels a strange mixture of fear and admiration. For EJ Marey the body is a machine that must be disassembled to be understood, with the help of other optical, mechanical or chemical machines, and finally partially re-built through other machines. One machine takes over from other machines (Sicard 2008; Chazal 2008). Such a symbiosis between Science and Art anticipates the recurrently promised fusion of the subject with the object, the body (the wet) with the machine (the dry). Fritz Lang's film *Metropolis*, shot with considerable resources in the mid-1920s in Berlin, is one of the first testimonies to the recurrent modern desire for a synthesis between the organic and the inorganic. The modern humanoid robot, which becomes alive fictitiously thanks to the photogenic computer graphic creatures of Hollywood film productions, is the modern version of such a fantasized syncretism. Today, the multitude of fantastic but cosmetic organisms present in visual space (e.g., *Mecha*, *Cyborgs*, *Clones*...) mostly contribute to a "mental morphing" that establishes an illusory continuity between a robot and a human being. Pr. Ishiguro who exhibits himself beside his metallic clone illustrates the aestheticisation of science, his "geminoïd" being an artifice supposed to facilitate human to robot interaction, and in so doing blurring the boundary between the dream and the rational. The scientific work that must seduce oscillates here between object and subject, materiality and aesthetics. Similarly, the ever-increasing weight of the image industry and the visual saturation it creates regularly inspire technoscience and sometimes indicate the direction in which rationality should go. This is ultimately a radical reversal of the initial situation: scientific photography as a source of improvisation for artists is today a medium to beautify science, a reversal that Marey the positivist would no doubt have not predicted. At last, the weight *of the movie* industry and the overconsumption of images are so heavy today that they inspire technoscience and determine the direction along which research should go. It is finally a radical reversal of a paradigm that initially inscribed scientific photography as a first source of improvisation for artists, but which now predominates and that Marey the positivist would not have predicted.

In sum, the possibility of making a recording of bodily action promoted a pictorial revolution and the constitution of a new language for both realistic and abstract schools of thought. Sketches of the human in motion thus strongly calibrated the occidental way of seeing, understanding and describing the living world. Muybridge and Marey's atlases became a key source of aesthetic modernism, something difficult to appreciate today. For our eyes have become so familiar with the serial display of body positions. As a result, a visual surprise is slowly being transformed into a pictorial convention. Finally, as soon as Science and Art started sharing the same pictorial medium, they moved closer and closer, and influenced each other even more. Even if Art and Science produce specific *atmospheres* (see Bollnow's 2011; Ingold 2015), a haze flowing forth from Marey's lab reached realist and abstract artworks while the scientific community could also breathe an artistic haze into the growing usage of scientific photography.

4.5   From Action to Action Perception

In addition to supplying human physiology with innovative tools, chronophotography, as a sensitive machine, opened up a new way to investigate human visual perception. Among the multiple uses the scientists will make of it, we should mention the Swedish psychologist Gunnar Johansson who underlined the influence of mental processes in perception. In his experiments, he elegantly demonstrated the role of the structure of the scene in the perception of an object in motion. Inspired by Marey's men in black and their dense compression of visual input, G. Johansson filmed objects or actors wearing small lights attached to each body joint. He thus highlighted the fact that the kinematics of gait now appeared isolated from the object or the body shape (the so-called point light animations or *PL* method, see Johansson 1973) since each dot is now moving inside a meaningless visual context. Further, the vertical translation at constant velocity of a PL between two other PLs engaged in translating horizontally and so forming a rigid structure is suddenly perceived as an elliptical motion when the structure disappears and the PL moves alone (e.g., see Johansson for an illustration). These experiments mark an important step in the advancement of knowledge about the visual system, which, after Johansson's experiments, is no longer seen as an isolated and peripheral physiological process localized at the level of the retina but as a cognitive process. Inspired by these seminal investigations, systematic studies will then describe the behavioral and neural responses associated with the perception of the actions of others (see Blake and Shiffrar 2007 for a review). More importantly, once it has become possible to study the visual perception of motion, psychologists and physiologists get closer to each other, and new questions could be raised about potential neural connections between perception and action systems. In the following part of this paper we present empirical data belonging to this theme of research that would not have been possible without the advent of chronophotography.

It is now accepted that visual sensitivity to human motion is partly related to the experience the observer accumulates during motor activities. Further, it is recurrently proposed that sensory inputs during action-perception reach both sensory and motor cortical areas (Rizzolatti and Craighero 2004). Based on these observations, we hypothesized that if human motion-perception relies on the transformation of seeing into doing, abnormal body motion should affect the neural processing of motion recognition. Following Johansson's PL method, we were able to test the role of the structure of the visual scene in the perception of human motion (Pozzo et al. 2017; Inuggi et al. 2018). More precisely, we checked whether pictorial information (the spatial location of the PLs) prevails, as compared to body motion information (the kinematics of the PLs), in the recognition of human motion, as it is the case for a lifeless object. We thus collected electroencephalographic (EEG) signals from participants during the observation of different human locomotor patterns manipulating the gestalt (by changing the position of the PLs initially located on body joints) and the motion of the PLs (by cancelling the net body forward translation). Concerning this later variable, most of previous studies investigated the perception of human locomotion by displaying PLs of walkers on a treadmill, this strange wandering consisting in walking in the reverse direction of an airport treadmill. We predicted that, if perception is kinematic dependent instead of body geometry dependent, a natural forward locomotion, in contrast to passive and reactive treadmill locomotion, would deliver the right kinematic feedback to the embodied visual processing. Indeed, natural forward locomotion suggests the walker's intention to reach a spatial goal (e.g., a bus station, or even something more improbable like a mushroom). On the contrary, artificial roaming on the spot is, for a viewer before the invention of the treadmill (e.g. a modern homo sapiens), an intransitive behavior from which it is difficult to divine the walker's intention. Finally, considering that an observer's motor experience (Viviani and Stucchi 1992; Calvo-Merino et al. 2006; Cannon et al. 2014; Quandt and Marshall 2014; Meirovitch et al. 2015) determines efficient interaction with conspecifics and also contributes to inferring an actor's intention successfully, the visual relevance of natural body translation becomes evident. Therefore four types of stimuli were created: a centered walker, a centered scrambled walker, (i.e., a scrambled form consisting of the display of side views of blobby shapes while PLs moved with the same amount of absolute motion, which however appear as a meaningless assemblage of PLs), a translating walker and a translating scrambled walker. Each stimulus was thus obtained by combining two factors with two levels each: shape (either walker or scrambled) and translation (that could be present or absent as during walking on a treadmill). We thus verified how these displays produced sensorimotor spectral perturbations in EEG signals. We hypothesized that if biological motion recognition is motor dependent, a significant difference could be expected as between translated compared to centered walkers with regard to their sensorimotor cortical activity. Our analysis revealed that translational component of locomotion induced greater motor resonance (a notion referring here to motor cortical activity induced by visual stimuli) than human shape. More precisely, translating compared to centered stimuli produced higher power decreases in the beta 1 and beta 2 bands (two key frequency bands whose desynchronization reflects a motor resonance process) in Superior Parietal areas. Interestingly, the desynchronization of beta band, even if reduced, was still present for translating scrambled stimuli and was reported by the

Video. 4.1   A two dots
walking man. See https://
drive.google.com/file/d/1wG
cyDlVpqDbvGZ1hxXAR
3Zz5GiklBznp/view?usp=
sharing

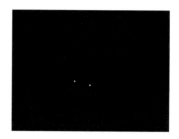

observers as a sort of "blob in motion". In fact, the scrambled display only modified the body structure whilst dot local motions remained compatible with motor representation, namely the *2/3 power law* (Lacquaniti et al. 1983; Viviani and Flash 1995; Flash and Hogan 1985; Flash and Handzel 2007; Bennequin et al. 2009), also present during treadmill locomotion (Ivanenko et al. 2002). For instance a cloud of dots without any recognizable gestalt but moving along elliptical trajectories strictly according to this motor rule is enough to activate dorsal premotor and supplementary motor areas (Dayan et al. 2007; Meirovitch et al. 2015).

In summary, a visual scene reduced to few moving dots brought significant sensorimotor information to activate the motor resonance process. This accords with the qualitative observation that human locomotion is still recognizable when viewing the trajectory of only two PLs located on the foot malleolus (see Video 4.1). Differently, the graphical trace left by the walker shows the life process in his totality. As soon as the white dots are animated the *perception of movement* becomes a *self-perception in movement*. In that case the *"movement"* (the perceived kinematics) would evoke intuitively the *"moving"* (the living object, see Merleau-Ponty 1960) and its corresponding but invisible body structure. One PL as an integral part of the body would thus be mentally attached to the PLs of the other body parts. In contrast, the motion of lifeless objects that change in the presence of additional surrounding PLs (as in Johansson experiment, see above) cannot be attached to one particular invariant configuration as the body geometry of a biped.

Overall, these results support the idea that the motor system plays a crucial role in visual perception and emotional recognition (see Barliya et al. 2013), especially when body geometry appears as a meaningless assemblage of dots, or when only one dot is displayed in the scene (Pozzo et al. 2006). Because the visible consequences of past behavior are usually poor or incomplete and the future is not already available, the missing feedback are compensated by internal signals generated by motor activities to reconstruct the past and then predict the future of the ongoing action. Where the perception is mediated by the motor system and multiple associated sensory predictions (Droulez and Berthoz 1990), the results disagree with the idea that visual esthetic experience is unimodal and relies exclusively upon the visual brain (Zeki and Lamb 1994).

4.6 Commonality and Singularity

Referring to Marey's work and the introduction of motion capture technique, we saw
that humans in motion progressively became a series of subtly discrete postural variations of one single action, which are then flattened out on a plane and finally plotted
onto lines. Marey's figurations, mapping points and trajectories, objectively fix what
the artist has designed subjectively, and pave the way for a quantitative behavioral
science, where all entities aggregated in a third-order metaphor (i.e., from shape
to lines and points) are equivalent. Ironically, after Marey's disembodied graphical
notation, artists who rejected scientific and realist artwork, will adopt geometric
body representations promoting an *art of traces* and showing commonalities. For the
tenants of the abstract school (e.g., see Kandinsky's or Kupka's artworks), a trace has
the beauty of simplicity, a reduction without which it becomes impossible to reach the
essence. This contrasts with the ostentatious art and mastery of the know-how, which
mainly emphasizes the skill of the artist instead of his thoughts and intentions. The
idea that abstraction can be defined as a simplification of the concrete, which through
a process of removing all inessential information arrives at its essentially universal
aspect has been recurrently proposed in psychology (see Zimmer 2003; Barsalou
2003). But this aesthetic approach, where the picture flees the world of appearances
and streamlines information processing, still poses problems for the neuroscientist
concerned by the role of the body in the emotional process: How a static graphical
trace can evoke motion and produce (embodied) aesthetic experience?

As we have sketched above, an important corpus of empirical data confirms that
perception is motor dependent, and that vision is sensitive to human motion, partly
because of concomitant cortical motor activities in the observer's brain. In this theoretical frame, a direct matching of the visual input to observer motor memory, via
the mirror neuron system (MNs), is recurrently proposed (Rizzolatti et al. 2001).
Following this neural pathway, MNs would allow the observer to put her/his feet
into the shoes of the actor (Jeannerod 2001). Therefore, watching a rough sketch of
human action would resonate with the observer's motor repertoire and this would
produce an aesthetic experience based on an embodied communication linking the
artwork to the spectator.

In an in-depth phenomenological analysis of Kandinsky's work, Michel Henry had
the intuition of such an automatic coupling between perception and action. Applying
a similar schema, he proposed that a line, like the trace of an elusive movement, is
*"…a pure sensation and a pure experience… that does not even need to translate,
through any means, the abstract content of our invisible life"* (Henry 1988, pp. 72).
Reworded, as soon as the abstract drawing empties the picture of its subject, it gains
in power of motor evocation. What is primordial here is not the recognition of an
object but the way the line has been performed. Further, traces omit information
that is irrelevant and may distract the observer, who now should be able to solve
the causal link between the trace and the organism that produced the trace. A poor
visual display would thus induce greater motor resonance than realistic copies of
the world. More hypothetically, lines and points would focus the gaze on traces,

inducing implicit motor inference as opposed to meaningful body shape that would distract from action observation. Thus, looking at a graphical trace left by a white marker stuck on Marey's man in black would help us to get inside him, and would establish a causal relation between the supposed beginning and end of the trace. An abstract trace, in contrast to a meaningful picture, would give us ways of expressing a biological preference for seeing things in motion rather just watching the things themselves. Finally, because it is extremely reduced in term of contour, contrast, shape and colors, a simple stroke would produce artwork in which gestures become visible, just as "Action-painting" made the painter's gestures visible by splashing the paint on the canvas. In the same vein, Freedberg and Gallese (2007) have explained abstract art in the perspective of embodied simulation. However, saying this implies that a keen observer should be able to perform the gesture instead of the artist, which imposes a strong constraint upon the embodied simulation hypothesis. Indeed, the human body consists of more than 600 muscles, far more than are necessary to perform our daily life activities, and one simple reaching or tracing movement can be performed in very different ways (see Hilt et al. 2016). Moreover, intersubjective variability is a decisive factor for self-building and agency. Indeed, motor styles determine the ability to perceive different gestalts, which, in turn, makes each of us unique and able to distinguish oneself from the other. But if each observed actor carries a motor style in discrepancy with the observer's own, a direct mapping of the external figure onto the observer's private motor representations turns out to be very tricky. In other words, how can a '*dividual*' (a divisible that can be shared by a number) actor and an *individual* (the smallest and indivisible unit which society can be reduced to) observer be matched?

Several investigations tested this idea, by assuming that action perception (through either audio or visual input) is automatically embodied through such a direct matching process, transforming the visual signal into a motor command, as during imitation. Recent results suggest that the so called "motor resonance" process recorded during action observation would reflect the perceived distance between the style of an actor and an observer, rather than reflecting the direct matching of actor and observer representations. Hence, in the speech domain, the degree of motor recruitment required for listening to syllable scales for the perceived distance between listener and speaker (Bartoli et al. 2015). In the same way, it was also shown that the corticobulbar excitability of lip muscles while listening to speech is greater for speech sounds that are far from the listener's motor repertoire (Schmitz et al. 2018). Consequently, the visual and auditory perception of action would not covertly simulate similarities but would estimate differences by contrasting oneself from the external visual/auditory model, the discrepancy between the two generating an error signal to the communication process.

4.7   Conclusion

Contemporary abstract art is undoubtedly indebted to the original visual formalism designed by Marey, notably his point light displays disconnected from any reality as pure lines of action reduced to their essentials. Recurrent speculations made in neuroesthetics suggest that abstract lines drawn on a canvas will figure out an invisible content that only becomes visible thanks to a motor resonance that implicitly matches the visual effect elicited by the behavioral cause producing the trace. But because a stroke of paint is meaningless and reveals neither the goal nor the intention of the draftsman, Marey's abstract 2D graphic traces, and later Kandinsky's lines, plunge the observer into the abyssal night of subjectivity, leaving him alone, faced with his own fantasies. In this frame, and if esthetic experience is conceived as an inter-subjective communication contrasting both actor and spectator singularities, an appropriate empirical neuroesthetic approach should start to analyze behavioral and neurocognitive styles, that is, the qualities in addition to the quantities. Oppositely, whilst Mareysian abstract figurations inspired many modern artists, the physiologist simultaneously contributed to elaborate one undefined subject (a norm) by collecting many individuals supposed to reflect a central tendency, the opposite of what could be expected from visual art, that is, the expression of idiosyncrasy. Consequently, neuro-esthetics seems condemned to shift its paradigms towards the investigation of singularity and to abandon the quantities in favor of the qualities.

   At last and to come back to the question that is at the core of the present book: do artists express or use brain geometries? For Poincaré, the word geometry is a meta construction of the intellect grounded in the sensorimotor experience of space. As he put it *"To localize an object simply means to represent to oneself the movements that would be necessary to reach it.....it is not a question of representing the movements themselves in space, but solely of representing to oneself the muscular sensations which accompany these movements and which do not presuppose the preexistence of the notion of space"* (Poincaré 1907, pp. 47). Accordingly, the word geometry actually refers to both a rational science and a by-product of the sensorimotor experience. As the word Arabesque means both a curvilinear trace and a dance movement, geometry is both a form that can be quantified offline and an action carried out in real time, the latter being the main means for the artist to express geometry.

Acknowledgements Most of the research related to this paper was carried out at the University of Ferrara and at Caps UBFC. We are very grateful to P Dominey and Christopher Macann for their close reading of an earlier version of this paper.

References

Azema, M. F., & Rivere, A. (2012). Animation in Paleolithic art: A pre-echo of cinema. *Antiquity, 86*, 316–324.

Barliya, A., Omlor, L., Ma, Giese, Berthoz, A., & Flash, T. (2013). Expression of emotion in the kinematics of locomotion. *Experimental Brain Research, 225*, 159–176.

Barsalou, L. W. (2003). Abstraction in perceptual symbol systems. *Philosophical Transactions of the Royal Society of London: Biological Sciences, 358*, 1177–1187.

Bartoli, E., D'Ausilio, A., Berry, J., Badino, L., Bever, T., & Fadiga, L. (2015). Listener-speaker perceived distance predicts the degree of motor contribution to speech perception. *Cerebral Cortex, 25*(2), 281–288.

Bennequin, D., Fuchs, R., Berthoz, A., Flash, T. (2009). Movement timing and invariance arise from several geometries. *Plos Computational Biology* (Vol. 5).

Bitbol, M. (2010). *De l'intérieur du monde: Pour une philosophie et une science des relations.* Paris: Flammarion.

Blake, R., & Shiffrar, M. (2007). Perception of human motion. *Annual Review of Psychology, 58*, 47–73. https://doi.org/10.1146/annurev.psych.57.102904.190152.

Bollnow, O. F. (2011). Human Space. In J. Kohlmaier (Ed.), (trans. C. Shuttleworth), London: Hyphen Press.

Bragaglia, A. G. (1913). *Fotodinamismo futurista.* Rome: Nalato Editore.

Braun, M. (1992). *Picturing time, The Work of Etienne-Jules Marey.* Chicago-London: The University of Chicago Press.

Calvo-Merino, B., Grèzes, J., Glaser, D. E., Passingham, R. E., & Haggard, P. (2006). Seeing or doing? Influence of visual and motor familiarity in action observation. *Current Biology, 16*, 1905–1910. https://doi.org/10.1016/j.cub.2006.07.065.

Cannon, E. N., Yoo, K. H., Vanderwert, R. E., Ferrari, P. F., Woodward, A. L., & Fox, N. A. (2014). Action experience, more than observation, influences mu rhythm desynchronization. *PLoS ONE, 9*, e92002. https://doi.org/10.1371/journal.pone.0092002.

Chazal, G. (2008). La question du temps à la lumière de la chronophotographie. In Marey, Penser le mouvement, L'harmattan, Paris.

Cutting, J. E. (2002). Representing motion in a static image: Constraints and parallels in art, science, and popular culture. *Perception, 31*, 1165–1193.

Dayan, E., Casile, A., Levit-Binnun, N., Giese, M. A., Hendler, T., & Flash, T. (2007). Neural representations of kinematic laws of motion: Evidence for action-perception coupling. *Processing National Academy of Sciences USA, 104*, 20582–20587. https://doi.org/10.1073/pnas.071003 3104.

Debord, G. (2002). *The society of spectacle.* Hobgoblin Press Canberra.

Didi-Hubermann, G., & Mannoni, L. (2004). Mouvements de l'Air. Etienne-Jules Marey, photographe des fluides. Paris, Gallimard/RMN.

Droulez, & Berthoz. (1990). The concept of dynamic memory in sensorimotor control. In Humphrey, & Freund (Eds) *Motor control: concepts and issues* (pp. 137–161). Chichester: Wiley and sons.

Elias, N. (2016). Humana condition. *EHESS, coll.* Paris: Audiographie.

Flash, T., & Hogan, K. (1985). The coordination of arm movements: An experimentally confirmed mathematical model. *Journal of Neuroscience, 5*(7), 1688–1703.

Flash, T., & Handzel, A. A. (2007). Affine differential geometry analysis of human arm movements. *Biological Cybernetics, 96*(6), 577–601.

Freedberg, D., & Gallese, V. (2007). Motion, emotion and empathy in esthetic experience. *Trends in Cognitive Sciences, 11*(5), 197–203.

Frizot, M. (2001). *Etienne-Jules Marey chronophotographe.* Paris: Nathan/Delpire.

Gilbreth, F. B., & Gilbreth, L. M. (1917). *Applied motion study: A collection of papers on the efficient method to industrial preparedness.* New York: Sturgis and Walton.

Henry, M. (1988). *Voir l'invisible, essai sur Kandinsky.* PUF, Paris: Bourin.

Hilt, P. M., Berret, B., Papaxanthis, C., Stapley, P., & Pozzo, T. (2016). Evidence for subjective values guiding posture and movement coordination in a free-endpoint whole-body reaching task. *Science Reports, 6,* 238–268.

Hoffmann, C. (2013). Superpositions: Ludwig Mach and Étienne-Jules Marey's studies in streamline photography. *Studies in History and Philosophy of Science, 44,* 1–11.

Ingold, T. (2015). *The Life of Lines.* Taylor & Francis Group: Routledge.

Inuggi, A., Campus, C., Vastano, R., Saunier, G., Keuroghlanian, A., & Pozzo, T. (2018). *Observation of point-light-walker locomotion induces motor resonance when explicitly represented.* An EEG Source Analysis Study: Front Psychol. https://doi.org/10.3389/fpsyg.2018.00303.

Ivanenko, Y. P., Grasso, R., Macellari, V., & Lacquaniti, F. (2002). Two-thirds power law in human locomotion: Role of ground contact forces. *NeuroReport, 13,* 1171–1174. https://doi.org/10.1097/00001756-200207020-00020.

Jeannerod, M. (2001). Neural simulation of action: A unifying mechanism for motor cognition. *Neuroimage, 14,* S103–S109.

Johansson, G. (1973). Visual perception of biological motion and a model for its analysis. *Perception and Psychophysics, 14,* 201–211. https://doi.org/10.3758/bf03212378.

Lacquaniti, F., Terzuolo, C., & Viviani, P. (1983). The law relating the kinematic and figural aspects of drawing movements. *Acta Psychological (Amst), 54,* 115–130. https://doi.org/10.1016/0001-6918(83)90027-6.

Mannoni, L. (1999). *Etienne-Jules Marey, la mémoire de l'œil.* Milan, Mazzotta: Cinémathèque française.

Marey, E. J. (1873). *La machine animale.* Paris: Germer Baillière.

Marey, E. J. (1884). *Développement de la méthode graphique par l'emploi de la photographie.* Paris: G Masson.

Marey, E. J. (1892). *La Photographie du mouvement.* Paris: Georges Carré.

Marey, E. J. (1899). *La chronophotographie.* Paris: Gauthier-Villars.

Meirovitch, Y., Harris, H., Dayan, E., Arieli, A., & Flash, T. (2015). a and b band event-related desynchronization reflects kinematic regularities. *Journal of Neuroscience, 35,* 1627–1637. https://doi.org/10.1523/JNEUROSCI.5371-13.2015.

Merleau-Ponty, M. (1960). *La structure du comportement.* Paris: PUF.

Poincaré, H. (1907). *The value of science.* New York: The Scientific Press.

Pozzo, T. (2003). La chronophotographie scientifique: aux origines du spectacle du monde vivant. In Semia (Ed.), Images, Science, Mouvement: autour de Marey. Paris: L'harmattan.

Pozzo, T., Papaxanthis, C., Petit, J. L., Schweighofer, N., & Stucchi, N. (2006). Kinematic features of movement tunes perception and action coupling. *Behavioural Brain Research, 169,* 75–82.

Pozzo. T. (1995a). De la chronophotographie à l'analyse moderne du mouvement: rupture ou continuité? In M. Leuba (Ed). *Marey, Pionners de la synthèse du mouvement* (pp. 61–69). Musée Marey: Beaune.

Pozzo, T. (1995). La chronophotographie: une approche moderne du mouvement humain. In J. Delimata (Ed.), *Marey/Muybridge, Rencontre Beaune/Stanford* (pp. 120–128). Conseil Régional de Bourgogne: Stanford University.

Pozzo, T. (2013). Physiologie de la vérité. La vérité, Colloque de l'IUF, édition EUSE, St Etienne.

Pozzo, T., Inuggi, A., Keuroghlanian, A., Panzeri, S., Saunier, G., & Campus, C. (2017). Natural translating locomotion modulates cortical activity at action observation. *Front System Neuroscience, 11,* 83. https://doi.org/10.3389/fnsys.

Quandt, L. C., & Marshall, P. J. (2014). The effect of action experience on sensorimotor EEG rhythms during action observation. *Neuropsychologia, 56,* 401–408. https://doi.org/10.1016/j.neuropsychologia.2014.02.015.

Rizzolatti, G., Fogassi, L., & Gallese, V. (2001). Neurophysiological mechanisms underlying the understanding and imitation of action. *Nature Reviews Neuroscience, 2,* 1–10.

Rizzolatti, G., & Craighero, L. (2004). The mirror-neuron system. *Annual Review of Neuroscience, 27*(1), 169–192.

Sadier, B., Delannoy, A., Benedetti, J. L., Bourles, D., Jaillet, S., Geneste, J. M., et al. (2012). Further constraints on the Chauvet cave artwork elaboration. *Proceedings of the National Academy of Sciences, 109,* 8002.

Schmitz, J. Bartoli, E., Maffongelli, L., Fadiga, L., Sebastian-Galles, N., & D'Ausilio, A. (2018). Motor cortex compensates for lack of sensory and motor experience during auditory speech perception. *Neuropsychologia* 1–7.

Seghier, M., Price, C. (2018). Interpreting and utilising Intersubject Variability in Brain Function. *Review Trends in Cognitive Sciences, 22*(6), 517–530.

Sicard, M. (2008). La précession de la trace. In *Marey, penser le mouvement.* Paris: L'harmattan.

Tufte, A. (1997). *Visual explanations: Images and quantities.* Evidence and Narrative: Graphics Press, Connecticut.

Tversky, B. (2015). The cognitive design of tools of thought. *Review Philosophical Psych, 6,* 99–116.

Viviani, P., & Stucchi, N. (1992). Biological movements look uniform: Evidence of motor-perceptual interactions. *Journal of Experimental Psychology: Human Perception and Performance, 18,* 603–623. https://doi.org/10.1037//0096-1523.18.3.603.

Viviani, P., & Flash, T. (1995). Minimum-jerk, two-thirds power law, and isochrony: Converging approaches to movement planning. *Journal of Experimental Psychology: Human Perception and Performance, 21,* 32–53.

Zeki, S., & Lamb, M. (1994). The neurology of kinetic art. *Brain, 117,* 607–636.

Zimmer, R. (2003). Abstraction in art with implications for perception. *Philosophical Transactions of the Royal Society of London: Biological Sciences, 358,* 1285–1291.

Chapter 5
A Neuro-Mathematical Model for Size and Context Related Illusions

B. Franceschiello, A. Sarti, and G. Citti

Abstract We provide here a mathematical model of size/context illusions, inspired by the functional architecture of the visual cortex. We first recall previous models of scale and orientation, in particular Sarti et al. in Biol Cybern 9:33–48, (2008), and simplify it, only considering the feature of scale. Then we recall the deformation model of illusion, introduced by Franceschiello et al. (J Math Imaging Vis 60:94–108, 2017b) to describe orientation related GOIs, and adapt it to size illusion. We finally apply the model to the Ebbinghaus and Delboeuf illusions, validating the results by comparing them with experimental data from Massaro and Anderson (J Exp Psychol 89:147, 1971) and Roberts et al. (Perception 34:847–856, 2005).

5.1 Introduction

Geometrical-optical illusions (GOIs) are a class of phenomena first discovered by German physicists and physiologists in the late XIX century, among them Oppel (1855) and Hering (1861), and can be defined as situations where a perceptual mismatch between the visual stimulus and its geometrical properties arise (Westheimer 2008). Those illusions are typically analyzed according to the main geometrical features of the stimulus, whether it is contours orientation, contrast, context influence, size or a combination of the above mentioned ones (Westheimer 2008; Ninio 2014; Eagleman 2001).

B. Franceschiello
Laboratory for Investigative Neurophysiology, Department of Radiology, Lausanne University Hospital and University of Lausanne (CHUV-UNIL), Lausanne, Switzerland

Department of Ophthalmology, Fondation Asile des aveugles and University of Lausanne, Lausanne, Switzerland

A. Sarti
CAMS-EHESS, Paris, France

G. Citti (✉)
Department of Mathematics, University of Bologna, Bologna, Italy
e-mail: giovanna.citti@unibo.it

© Springer Nature Switzerland AG 2021   91
T. Flash and A. Berthoz (eds.), *Space-Time Geometries for Motion and Perception in the Brain and the Arts*, Lecture Notes in Morphogenesis,
https://doi.org/10.1007/978-3-030-57227-3_5

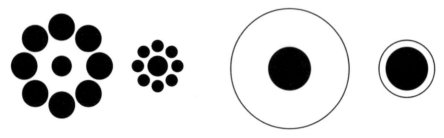

Fig. 5.1 The Ebbinghaus illusion (left) and the Delboeuf illusion (right)

In this work we are mainly interested in size and context related phenomena, a class of stimuli where the size of the surroundings elements induces a misperception of the central target width. In Fig. 5.1, two famous effects are presented, the Ebbinghaus and Delboeuf illusions: the presence of circular inducers (Fig. 5.1, left) and of an annulus (Fig. 5.1, right) varies the perceived sizes of the central targets. These phenomena have been named after their discoverers, the German psychologist Hermann Ebbinghaus (1850–1909), and the Belgian philosopher and mathematician Joseph Remi Leopold Delboeuf (1831–1896), Delboeuf (1865). The Ebbinghaus phenomenon has been popularized in the English-speaking world by Edward B. Titchener in a textbook of experimental psychology 1991, and this is the reason why it is also called Titchener illusion (Roberts 2005).

The importance of studying these phenomena at a psychophysical and neuroimaging level lies in the fact that these phenomena provide insights about the functionality of the visual system (Eagleman 2001). Many studies show that neurons of at least two of the visual areas, V1 and V2, carry signals related to illusory contours, and that signal in V2 is more robust than in V1 Von Der Heyclt et al. (1984), Murray et al. (2002), reviews Eagleman (2001), Murray and Herrmann (2013), see Fig. 5.2 from Murray and Herrmann (2013). As for what concerns size and context-dependent phenomena such as those presented in Fig. 5.1, it is not new that attention plays a huge role in modulating the visual response (Yan et al. 2014). A further proof lies in the usage of these context dependent illusions for proving different perceptual mechanisms related to attention, in cross-cultural study, see (Doherty et al. 2008, Bremner et al. 2016, Fonteneau et al. 2008).

Geometrical models for optical illusions related to orientation perception were proposed by the pioneering work of Hoffman (1971), in term of Lie groups, and then by Smith (1978), who stated that the apparent curve of GOIs (where the main feature is orientation) can be modeled by a first-order differential equation. A first attempt was performed also by Walker (1973), who tried to combine neural theory of receptive field excitation together with mathematical tools to explain misperception of straight lines in GOIs. These results, together with Smith (1978) and Ehm and Wackerman in (2012), introduced a quantitative analysis of the perceived distortion. On the other hand another possible way to go is to use a Bayesian approach to model the neural activity, approach who inspired to Helmholtz's theory (von Helmholtz and

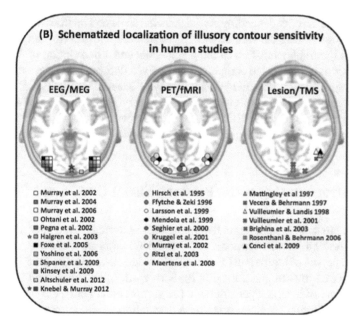

Fig. 5.2 Quoting from Murray and Herrmann (2013), copyright licence number 4936950076901: (B) Schematic localization of Illusory contours (IC) sensitivity in human studies. The colored symbols indicate the approximate locations of IC sensitivity for human studies using electroencephalography (EEG)/magnetoencephalography (MEG) source estimations (left), positron emission tomography (PET) and functional magnetic resonance imaging (fMRI) (middle), and lesion studies or transcranial magnetic stimulation (TMS) (right). The stars in the left panel indicate secondary and subsequent effects

Southall 2005; Geisler and Kersten 2002). These methods allowed to consider how prior experience influences perception, (Knill and Richards 1996), and were applied to motion illusions by Weiss et al. in (2002). Fermüller and Malm in (2004) attributed the perception of geometric optical illusions to the statistics of visual computations. More recently the authors in Franceschiello et al. (2017a, 2017b) proposed a model for orientation-based geometrical illusions inspired by the functionality of simple cells of the visual cortex (Citti and Sarti 2006). Geometric models of the functionality of the visual cortex were proposed by Hoffman (1989), Mumford in (1994), Williams and Jacobs in (1997), and more recently by Petitot and Tondut in (1999) and Citti and Sarti in (2006). These models were mainly focused on orientation selectivity, but they have also been extended to describe scale selectivity in Sarti et al. (2008, 2009).

The aim of this paper is to extend the work in Franceschiello et al. (2017a, 2017b), to illusion of size and scale, starting from the cortical model of Sarti et al. (2009). In their models families of simple cells are characterized by a cortical connectivity and a functional geometry. The main idea is that the context modulates the connectivity metric and induces a deformation of the space, from which it will be possible to compute the displacement and the corresponding perceived misperception. An isotropic functional connectivity depending on the detected scale and on the distance between the objects composing the stimuli will be considered. It will follow an explanation concerning the implementation of the phenomena and a description of the numerical simulations performed to compute the perceived deformation. The computations will be in agreement with a judgemental study of Massaro et al. (1971), as well as with the observations of how illusions change by varying the distance between target and inducers, (Roberts et al. 2005). To our best knowledge, this is the first original contribution providing an interpretation to size related geometrical optical illusions.

5.2 Neurogeometry of the Primary Visual Cortex and GOIs

Neuromathematical models target features encoding during early stages of the visual process. The first geometric models of the functionality of the visual cortex date back to the papers of Hoffmann (1989) and Koenderink-van Doorn (1987). Citti and Sarti developed in Citti and Sarti (2006), Sarti et al. (2008), Sarti and Citti (2015, 2015), a theory of invariant perception in Lie groups, taking into account (separately or together) different features: brightness orientation, scale, curvature, movement. Other papers applying instruments of Lie groups and differential geometry for the description of visual processing have been introduced by August and Zucker (2000), Petitot and Tondut (1999), Duits and Franken (2010a, b).

5.2.1 The Receptive Field of a Cortical Neuron

The visual process is the result of several retinal and cortical mechanisms acting on the visual signal. The retina is the first part of the visual system responsible for the transmission of the signal, which passes through the Lateral Geniculate Nucleus, where a first preprocessing is performed, and arrives in the visual cortex, where it is further processed. The receptive field (RF) of a cortical neuron is the portion of the retina which the neuron reacts to, and the receptive profile (RP) $\psi(\xi)$ is the

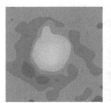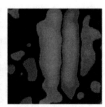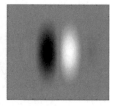

Fig. 5.3 From left to right: in vivo registered radial symmetric receptive fields, see De Angelis et al. in (1995); their model as Laplacian of a Gaussian; in vivo recorded odd receptive field (from De Angelis et al. in (1995)); their model as Gabor filter, see (5.1)

function that models the activation of a cortical neuron when a stimulus is applied to a point $\xi = (\xi_1, \xi_2)$ of the retinal plane. As an example, we recall that the RP of simple cells sensible of scale and orientation have been experimentally described by De Angelis in (1995), and modelled as a Gabor filter in Daugman (1985), Jones and Palmer (1987), see Fig. 5.3. If T_{x_1, x_2} is a translation of vector (x_1, x_2), D_σ a dilation of amplitude σ and R_θ is a rotation of an angle θ, a good expression for the Gabor filters sensible to position (x_1, x_2), orientation and scale (θ, σ), is:

$$\psi_{x_1, x_2, \sigma, \theta}(\xi) = D_\sigma R_\theta \psi_0 T_{x_1, x_2}(\xi), \quad \text{where} \quad \psi_0(\xi) = \frac{1}{4\pi} e^{-\frac{4\xi_1^2 + \xi_2^2}{8}} e^{2i\bar{b}\xi_2}. \tag{5.1}$$

5.2.2 Output of Receptive Profiles

Due to the retinotopic structure, there is an isomorphism between the retinal and cortical plane in V1, which we will discard in first approximation. Furthermore the hypercolumnar structure, discovered by the neurophysiologists Hubel and Wiesel in the 60s (1977), organizes the cells of V1/V2 in columns (called hypercolumns), each one covering a small part of the visual field $M \subset \mathbb{R}^2$ and corresponding to parameters such as orientation, scale, direction of movement, color, for a fixed retinal position (x_1, x_2). Over each retinal point we will consider a whole hypercolumn of cells, each one sensitive to a specific instance of the considered feature f, see Fig. 5.4. We will then identify cells in the cortex by the three parameters (x_1, x_2, f), where (x_1, x_2) represents the position of the point and f is a vector of extracted features. We will denote with F the set of features, and consequently the cortical space will be identified as $R^2 \times F$.

The retinal plane is identified with the $\mathbb{R}^2$-plane, whose local coordinates will be denoted with $x = (x_1, x_2)$. When a visual stimulus I of intensity $I(x_1, x_2) : M \subset \mathbb{R}^2 \to \mathbb{R}^+$ activates the retinal layer of photoreceptors, the neurons whose RFs intersect M spike and their spike frequencies $O(x_1, x_2, f)$ can be modeled (taking into account just linear contributions) as the integral of the signal $I(x_1, x_2)$ with the set of Gabor filters. The expression for this output is:

$$O(x_1, x_2, f) = \int_M I(\xi_1, \xi_2) \, \psi_{(x_1, x_2, f)} (\xi_1, \xi_2) \, d\xi_1 d\xi_2. \tag{5.2}$$

5.2.3 Cortical Connectivity

Note that the output is a higher dimensional function, defined on the cortical space. The lateral connectivity propagates this output in the cortical space $R^2 \times F$ giving rise to the cortical activity. This corresponds to a localized activity in the cortex, where

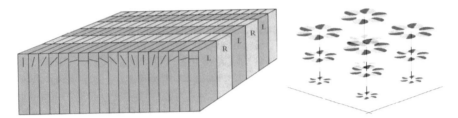

Fig. 5.4 Left: representation of the hypercolumnar structure, for the orientation parameter, where L and R represent the ocular dominance columns (Petitot 2008). Right: for each retinal position (x_1, x_2), according to the model in Sarti et al. (2009), there is the set of all possible orientations and scales

neurons are parametrized by the variables (x, f). Interactions between synaptically coupled neurons occur via events called action potentials. A single action potential evokes a voltage change (post synaptic potential, PSP) in the postsynaptic element. Cortical connectivity has been measured in many families of cells, and it is strongly anisotropic. It has been proved that there is a relation between the shape of the receptive profiles, their connectivity and their functionality. A good model for the cortical connectivity can be obtained describing $R^2 \times F$ as a Lie group, endowed with a sub-Riemannian metric, or a symplectic structure, and we here define K_F as a decreasing exponential function of the distance. Typically a fundamental solution of a Fokker Planck equation, left invariant with respect to the group law has this property. Since we are interested into scale models, we recall here the model of Sarti et al. (2008) and Petitot in (2008) who proposed a model of scale and orientation selectivity in $R^2 \times S^1 \times R^+$, with the group low of translation, rotation and dilation. A basis of left invariant vector fields can be defined as:

$$X_1 = \sigma(\cos\theta \partial_{x_1} + \sin\theta \partial_{x_2}); \quad X_2 = \sigma \partial_\theta;$$
$$X_3 = \sigma^2(-\sin\theta \partial_{x_1} + \cos\theta \partial_{x_2}); \quad X_4 = -\sigma^2 \partial_\sigma.$$

A geometrical structure compatible with the observed connectivity is the Riemannian metric g_F which makes the vector fields X_1, X_2, X_3, X_4 orthonormal. Indeed, due to the different scale factor, X_1/σ, X_2/σ have a non zero limit, when σ goes to 0, while X_3/σ, X_4/σ tend to 0. For this reason, this metric couples naturally the vector fields X_1, X_2 and X_3, X_4. The integral curves starting from a fixed point of the first two vectors and the second two vectors give rise to two families of curves (see Fig. 5.5, top) whose 2D projection reveals the same pattern of co-axial and trans-axial connections measured by Yen (1998), validating the model at a neuro-physiological level (see 5.5, bottom). Therefore, a decreasing function of the distance can be considered a good model for the connectivity kernel.

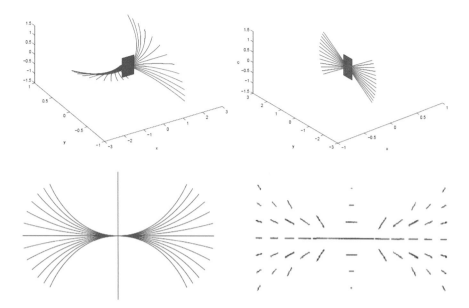

Fig. 5.5 Top left: integral curves of the vector fields X_1 and X_2 (blue). Top right: integral curves of the vector fields X_3 and X_4 (red). Bottom left: their 2D projection. Bottom right: this projection has patters compatible with the measured connectivity patterns [from Yen (1998)]

5.2.3.1   Non-maxima Suppression

The cortex is equipped with an intracortical neural circuitry which acts within a single hypercolumn. In presence of a visual stimulus, at a point $x = (x_1, x_2)$, the whole hypercolumn over that point fires, but mechanisms of non-maximal suppression act, suppressing the output of cells that within the same hypercolumn are not maximally firing. In this way the connection is able to sharpen the tuning of feature selection over each point $x = (x_1, x_2)$. This selection defines a value of the feature f (scale) at every point. The points of maximal response will be denoted from now on as $\bar{f}(x)$. The output will then have the following expression:

$$O(x, \bar{f}(x)) = \max_f O(x, f) \tag{5.3}$$

5.2.3.2   Long Range Connectivity

If a family of cells has been described via $R^2 \times F$, with a metric g_F and a connectivity kernel K_F, the problem is to describe the connectivity action which induces the cortical activity. It can be described through a mean field equation, following an approach first proposed by Wilson and Cowan (1972), Amari (1972), Ermentraut-Cowan (1979), Bressloff and Cowan (2003), to quote a few. The equation in its general formulation as the following expression:

$$\frac{\partial}{\partial t}a(x,t) = -a(x,t) + \int K_F(x-x',f-f)\psi\Big(a(x',f',t') + O(x',f') - C\Big)\,dx'df'$$

(5.4)

where ψ is a sigmoid, α and β suitable constants, C is a normalization factor. The equation can be applied in the lifted space or projected in the 2D space, via the non maxima suppression mechanism. The associated stationary equation satisfies

$$a(x,t) = \int K_F(x-x',f-f)\psi\Big(a(x',f',t') + O(x',f') - C\Big)\,dx'df' \qquad (5.5)$$

5.2.4 A Model for GOIs Related to the Orientation

In Franceschiello et al. (2017a, 2017b) two main ideas are developed for explaining orientation related GOIs. The initial stimulus is able to modulate the functional geometry of V1 and the geometry induced by the background of the perceived image can induce a perceptual deformation.

5.2.4.1 The Metric Modulated by the Visual Stimulus

The main idea developed in Franceschiello et al. (2017a, 2017b) is to modify the model for the functional geometry of V1 provided in Citti and Sarti (2006) and to consider that the image stimulus will modulate the connectivity: the new metric will be expressed as

$$||a(x,f)||g_F(x,f).$$

When projected onto the visual space, the modulated connectivity gives rise to a Riemannian metric which is at the origin of the visual space deformation. In the isotropic case, $g_F(x,f) = Id$ is the identity, and the metric reduces to a single positive real value. In this case non maximal suppression within the hypercolumnar structure is sufficient for explaining the mechanism, hence the metric induced on the 2D plane will be simply computed as:

$$||a(x,f(x))||Id.$$

If at every point we consider a non isotrotropic metric g_F, the projection is obtained by an integration along the fiber F at the point x. We refer to Franceschiello et al. (2017b) where the idea is discussed in detail.

5.2.4.2 Retrieving the Displacement Vector Fields

The mathematical question is how to reconstruct the displacement starting from the strain tensor $\mathbf{p}$. We think at the deformation induced by a geometrical optical illusion as an isometry between the $\mathbb{R}^2$ plane equipped with the metric $\mathbf{p}$ and the $\mathbb{R}^2$ plane with the Euclidean metric $\mathbf{Id}$:

$$\Phi : (\mathbb{R}^2, \mathbf{p}) \rightarrow (\mathbb{R}^2, \mathbf{Id}).$$

In strain theory $\mathbf{p}$ is called *right Cauchy-Green tensor* associated to the deformation Φ, for references see Lubliner (2008), Marsden and Hughes (1994). It is clear that it is equivalent to find Φ or the displacement as a map

$$\bar{u}(x_1, x_2) = \Phi(x_1, x_2) - (x_1, x_2),$$

where $(x_1, x_2) \in \mathbb{R}^2$. We can now express the right Cauchy-Green tensor in terms of displacement u. For *infinitesimal deformations* of a continuum body, in which the displacement gradient is small ($\|\nabla \bar{u}\| \ll 1$), it is possible to perform a geometric linearization of strain tensor introduced before, in which the non-linear second order terms are neglected. Under this assumption it was proved in Franceschiello et al. (2017b) that u is a solution of the PDE system:

$$\begin{cases} \Delta u_1 = \frac{\partial}{\partial x_1}p_{11} + 2\frac{\partial}{\partial x_2}p_{12} - \frac{\partial^2}{\partial x_1 \partial x_2}u_2 & \text{in } M \\[2mm] \Delta u_2 = \frac{\partial}{\partial x_2}p_{22} + 2\frac{\partial}{\partial x_1}p_{12} - \frac{\partial^2}{\partial x_1 \partial x_2}u_1 & \\[2mm] \frac{\partial}{\partial \mathbf{n}}u_1 = 0 & \text{in } \partial M \\[2mm] \frac{\partial}{\partial \mathbf{n}}u_2 = 0 & \end{cases} \qquad (5.6)$$

where M is an open subset of $\mathbb{R}^2$ and ∂M is Lipschitz continuous, with normal defined almost everywhere. Solutions for Eq. (5.6) are well defined up to an additive constant, which is recovered imposing $u(0, 0) = v(0, 0) = 0$ for symmetry reasons, where $(0, 0)$ is the center of our initial domain M.

5.3 The Model for Scale/Size GOIs

In this section we develop a model for scale type illusory phenomena. Once the connectivity is described, it will be used to define a new strain metric tensor. This enable us to adapt the model presented in Sect. 5.2.4.2 to this new features space and to recover the displacement vector fields induced by size perception.

5.3.1   Scale and Size of an Object

It is well known that simple cells of V1 are able to select the scale of an object, which is approximately the distance from the boundary [see for example Sarti et al. (2008), Petitot (2008)]. Strictly related to the scale is the size of the object, which represents the spatial dimension of the observed element. The Ebbinghaus and Delboeuf illusions (Fig. 5.1) are phenomena in which the context induces a misperception of the size of the central target, (Künnaps 1955). As a first size-evaluation of the perceptual units in an image is performed at early stages of the visual process, it is possible to adapt the cortical model to introduce a mechanism of non-maximal suppression able to evaluate the scale within an object. Therefore we introduce a metric in the position-size space. Following the intuition that there is a relation between the functionality cells and the shape of the connectivity, we assume that the connectivity related to scale and size values, which are real quantities, is isotropic. As a result the connectivity will decrease with the euclidean distance between the objects of the image. Finally we adapt the displacement algorithm to this metric, as recalled in Sect. 2.6, in order to model the illusion in Fig. 5.1.

5.3.2   Scale Selection in V1

In the previous section we recalled the model of orientation and scale selection of Sarti et al. (2008) and Petitot (2008). Here we will discard the orientation selection and we focus on the scale detection only. Scale is an isotropic feature, and can be selected by isotropic cells, as for example mexican hat cells, measured by De Angelis (see Fig. 5.3) A good model for their receptive profiles are the Laplacian of Gaussian. Denoting T_{x_1,x_2} a translation of a vector (x_1, x_2) and D_σ the dilation of amplitude σ, the bank of filters is represented as

$$\psi_{x_1,x_2,\sigma} = D_\sigma T x_1, x_2(\psi_0), \text{ where } \psi_0 = \Delta G, \text{ and } G(\xi_1, \xi_2) = \frac{1}{\pi} e^{-(\xi_1^2 + \xi_2^2)}.$$

The set of profile is then parametrized by the variables (x_1, x_2, σ), where (x_1, x_2) is the spatial position and σ is the scale variable. The bank of filters acts on the initial stimulus and the hypercolumns response of simple cells provides an output as the feature varies: this mechanism is described in Eq. (5.2). In the case of having only the scale feature $f = \sigma$ involved, the output reduces to:

$$O(x_1, x_2, \sigma) = \int_M I(\xi_1, \xi_2) \, \psi_{(x_1, x_2, \sigma)} (\xi_1, \xi_2) \, d\xi_1 d\xi_2. \tag{5.7}$$

The intra cortical mechanism selects the maxima over the orientation and scale hypercolumns, providing the selection of two maximal outputs for both features: $\bar{\sigma}$,

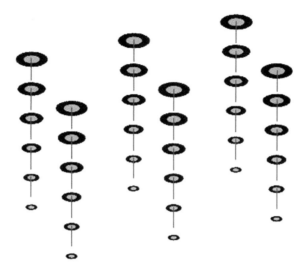

Fig. 5.6 The hypercolumnar structure for the scale model

as described in Eq. (5.3):

$$O(x, \bar{\sigma}(x)) = \max_{\sigma} O(x, \sigma) \qquad (5.8)$$

The maximum scale value $\bar{\sigma}$ represents the distance from the nearest boundary, selected over the hypercolumns containing all the possible distances σ. This is visualized in Fig. 5.6, where a bank of filter with different scales, but same orientation, is superimposed to a gray circle (the visual stimulus): the best fit is realized by the central image, whose scale is equal to the distance from the boundary. In Fig. 5.7, left we visualize an initial stimulus, the illusion and apply the scale selectivity maximisation (Fig. 5.7, right). The level lines of the function $\bar{\sigma}$ are circles, which describe the distance from the boundaries.

5.3.3 Size Selection

Once the distance function from the boundary $\bar{\sigma}(x_1, x_2)$ has been defined, we assume that the action of the connectivity propagates the output within each perceptual unit. Since the size $\rho(x_1, x_2)$ of an object can be identified as the maximum distance from the boundary, we postulate the action of a new non maxima suppression procedure, which takes place within the perceptual unit. This can be implemented through an advection equation

$$\frac{\partial \rho(x_1, x_2)}{\partial t} = |\nabla \bar{\sigma}(x_1, x_2)| \qquad (5.9)$$

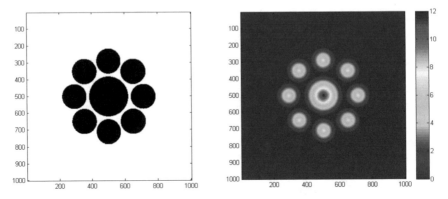

Fig. 5.7 Left: the initial stimulus processed. Right: the maximum response $\bar{\sigma}$. For each point the color identifies the distance $\bar{\sigma}$ from the nearest boundary

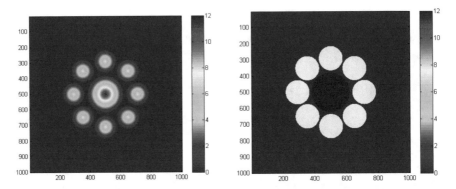

Fig. 5.8 Left: representation of $\bar{\sigma}(x_1, x_2)$. Right: propagation of the information within each circle using an advection equation. This allows us to recover for each perceptual unit the corresponding value of size, $\rho(x_1, x_2)$

This describes a conservation law which associates a single size value $\rho(x_1, x_2)$ to each perceptual unit of the image.

This step of the algorithm is visualized in Fig. 5.8. Starting from the left map representing the value of $\bar{\sigma}(x_1, x_2)$ previously detected, we propagated the maximum distance from the boundary within each circle using an advective equation, see (5.9).

5.3.4 Cortical Connectivity for Scale Type Illusion

Here we introduce the isotropic connectivity accounting for the interaction of points in scale illusions. As we mentioned in Sect. 5.2.3, we postulate here a strong relation between the functionality of the cells and shape of connectivity. Here the set of filters is generated by a fixed one by translation and dilation. Since D_σ and Tx_1, x_2

commute, then the set (x_1, x_2, σ), is a commutative group, and we can consider an isotropic metric on the space. Consequently the g_F will simply be the identity. In analogy, we will consider an isotropic metric also in the size space. As a consequence, the connectivity kernel will be an exponential decaying function of the Euclidean distance among objects composing the stimulus:

$$K_F(x - x') = \exp^{-c\,|x-x'|} \tag{5.10}$$

The long range spatial interaction decays when the spatial distance of cells increases. Here the kernel is an exponential, but it can be modeled more in general as a function decreasing with the distance. In analogy with the Bressloff- Cowan activity equation recalled in (5.5), the stationary activity equation will be expressed as the product between a connectivity kernel and the computed sizes of the objects:

$$a(x) = \int_{\mathbb{R}^2} \exp^{-c\,|x-x'|}(a(x') - \rho_0)\, dx' \tag{5.11}$$

where ρ_0 is a global normalization term denoting the *effective size*. It is a mean value for the activity. Since we are interested in evaluating the deformation of the target, we will choose $\rho_0 = \rho(0)$, so that ρ_0 represents the effective size of the central target.

5.3.5 Displacement Vector Field

The Euclidean metric, endowing the visual space, will be modulated by stimulus through the cortical activity computed above. Therefore, the metric induced by the stimulus in the 2D retinal plane will become:

$$p = a(x_1, x_2)\mathbf{Id}$$

at every point. The final step is to adapt the displacement equation to the present setting at every point of the space, and according to Sect. 5.2.4.2 we now look for the displacement map

$$u = \Phi - \mathbf{Id}, \quad \text{where} \quad \Phi : (\mathbb{R}^2, \mathbf{p}) \rightarrow (\mathbb{R}^2, \mathbf{Id}). \tag{5.12}$$

In other word Φ is the deformation which sends the metric p in the identity at every point. Due to the particular structure of the metric p, the equation that in (5.6) provides the value of u simplifies, as the coefficients p_{12} identically vanish. Indeed in this case the equation for u expresses a Cauchy Riemann—type condition. This means that the solution is harmonic and different from 0 and results into a radial vector field.

5.4   Implementation and Results

In this section the implementation of the presented model is presented and the results discussed. At a first glance, the discrete version of Eq. (5.11) becomes:

$$a(x) = \sum_{i=1}^{N} \exp^{-|x-x^i|}(\rho(x^i) - \rho_0) \qquad (5.13)$$

By simplicity, since ρ and a are locally constant, we assume that N is the number of inducers, and x^i, $\forall i$, represents the point in the inducer where the scale is maximal and coincides with the size. Another approximation consists in assuming that the constant c in the exponential map is 1, meaning $c = 1$. The distance $|x - x'|$ is expressed in pixels, $\rho(x')$ is the size of the inducer at point $x' = (x'_1, x'_2)$. We always consider points of the image in which the maximum of the scale is attained. The differential problem in (5.6) is then approximated with a central finite difference scheme and it is solved with a classical PDE linear solver.

5.4.1   Ebbinghaus Illusion

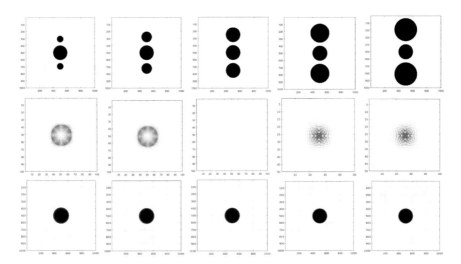

Fig. 5.9 First row: five Ebbinghaus illusion with two inducers of increasing width. Second row: the associated displacement vector fields are visualized. Third row the deformation of the target is visualized in black (the reference circle is visualized in red). If the inducers are smaller than the target, this expands, if the inducers are larger, the target shrinks

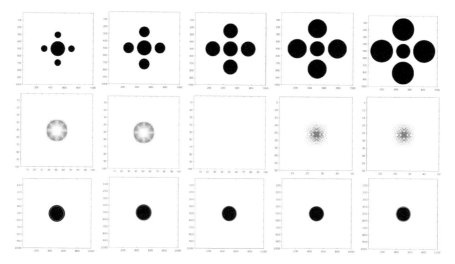

Fig. 5.10 Five Ebbinghaus illusion with four inducers (first row), the corresponding displacement vector field (second row) and the deformation of the central target (third row)

We will perform a first test on the Ebbinghaus illusion (Fig. 5.1, left). This illusion consists in central circle—target—surrounded by a number of circles—inducers. The *perceived size of the target*, which is the perceptual component we want to evaluate in this study varies if the size of the inducers varies (Massaro and Anderson 1971; Roberts et al. (2005) and if the distance between the inducers and the target increases or decreases Roberts et al. (2005). It has been experimentally if that whether the inducers are smaller than the target, the latter is perceived larger than its actual size. If the dimension of the inducers increases (but remains smaller than the target), the strength of the effect decreases, until the dimension of the target and the inducers are the same. In this last scenario the perceived dimension and the real dimension of the target coincide. If the dimension of the inducers increase, the target is perceived as if it was smaller. This happens independently of the number of the inducers (see Figs. 5.9, 5.10 and 5.11).

5.4.1.1   Quantitative Results: Changing the Number of Inducers

A quantitative analysis of the phenomenon has been made by Massaro and Anderson (1971). In a first experiment, they considered a target circle with diameters of 13 or 17 mm. There were two, four, or six context circles, symmetrically located around the target. The diameters of the context circles differed from the center circle by 8, 4, 0, −4, or −8 mm, for each size of central circle. The distance between the proximal edges of the central and context circles was always 6 mm. The figures were presented in four separately randomized blocks each day, for 2 days, to 10 subjects. They were instructed to judge the apparent size of the target. Responses

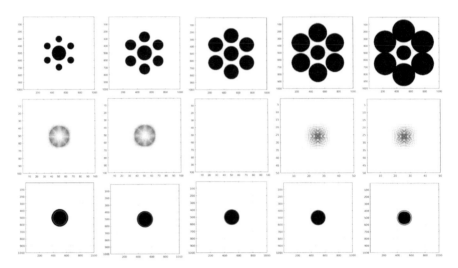

Fig. 5.11 The Ebbinghaus illusion with six inducers, and increasing inducers (first row), the computed displacement vector field (second row) and the deformation of the central target (third row)

were made by rotating a wheel that presented a single comparison circle. Figure 5.12, left, summarises the experimental results. In the abscissa the number of surrounding circles is represented, in the ordinate the judged size. If the size of the inducers is fixed, the perceived size of the target grows linearly with the number of inducers. To validate the model, we started from the same values used in Massaro and Anderson (1971) as for target measure, number of inducers and distance between target and inducers. The diameter of the target wss chosen equal to $\rho_0 = 14.6$, the number of inducers was $N = 2, 4, 6$ respectively and the size of the inducers was varied as follows: $\rho(x') = \rho_0 - 8, \rho_0 - 4, \rho_0, \rho_0 + 4, \rho_0 + 8$ pixels. Moreover the distance $|x - x'| = 6$ between target and inducers was kept fixed. The resulting Ebbinghaus images are depicted on the first row of Figs. 5.9, 5.10 and 5.11. In the second row the displacement computed through the infinitesimal strain theory approach is drawn. Finally, the third row contains the perceived central target in black. The red circle is the target reference of the initial stimulus, drawn in order to allow a comparison between the proximal stimulus (displaced image) and the distal one (physical stimulus). Figure 5.12, right, summarises the results found with our model, formally organized as in Massaro and Anderson (1971). The modeled and experimental results correctly match.

5.4.1.2 Quantitative Results: Changing the Distance Between Target and Inducers

In a second experiment Massaro and Anderson (1971) considered a family of Ebbinghaus illusion with six context circles. The diameter of the center circle was 13 mm,

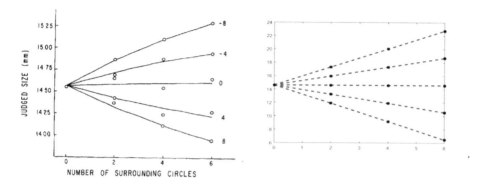

Fig. 5.12 Left: Massaro and Anderson (1971) shown that the perceived size of the central target varies in relationship with the size and the numbers of inducers. Right: the experimental outcome are exactly reproduced by our model

the diameter of the context circles was 5 mm. The distances between the proximal edges of center and context circles were 3, 6, 12, and 24 mm respectively. The stimuli were presented to 24 subjects six times in successive randomized blocks. The results

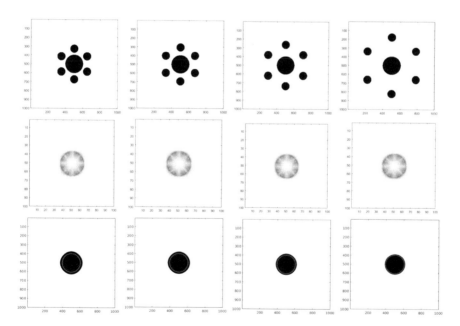

Fig. 5.13 Four Ebbinghaus illusions with increasing distance between the target and the inducers (first row). In the second row the computed displacement vector fields, and in the third row the computed deformation of the central target. If the distance is small, the target expands, while increasing the distance induces a perceptual shrink of the central target, correctly reproduced by the model

are collected in Fig. 5.14: the distance from the center circle is represented in the x-axis, and the judged size decrease non linearly with the distance.

We repeated the same experiment with our model. It is represented in Fig. 5.14. In the first row the four Ebbinghaus illusions, in the second row the computed displacement vector fields, and in the third row the modelled deformation of the central target. Finally in Fig. 5.14 right, we graphically represent the computed outcome in order to perform a comparison with the experimental one. It is easy to check that they correctly match, Fig. 5.15.

5.4.2 Delboeuf Illusion

The Delboeuf illusion, consists in a central black circle (the target) surrounded by an annulus, whose presence induces a misperception of the target size (see Fig. 5.16, first row). If the annulus is big, the target tends to shrink. If it is small, the target is perceived as expanding.

We apply the presented model to this illusion. Formula (5.13) becomes in this setting:

$$\rho(x) = \exp^{-|x-x'|}(\rho(x') - \rho_0) \tag{5.14}$$

where N of formula (5.13) is equal to 1, because the inducer is the annulus, $c = 1$, the distance $|x - x'|$ is the distance between the center of the target x and the center of the annulus x', expressed in pixels; $\rho(x')$ is the size of the annulus and ρ_0 refers again to the effective size. Then $(\rho(x') - \rho_0)$ expresses the difference between the considered sizes.

5.4.2.1 Discussion of the Results

We validate the model comparing our computational results with the experimental ones in Roberts et al. (2005). In their experiment, the authors focused on the effect

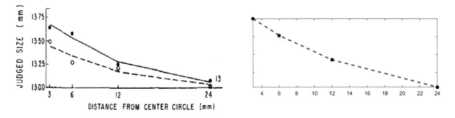

Fig. 5.14 Left: Massaro and Anderson (1971) shown how the perceived size of the central target decreases with the distance between the inducers and the target in the Ebbinghaus illusion. Right: we reproduce the experimental results

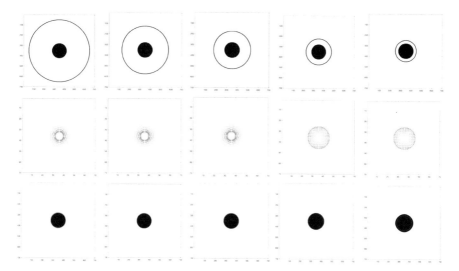

Fig. 5.15 First row: five Delbouef illusion with decreasing width for the annulus from left to right. Second row: the associated displacement vector fields are visualized. Third row the deformation of the target is visualized in black (the reference circle is visualized in red)

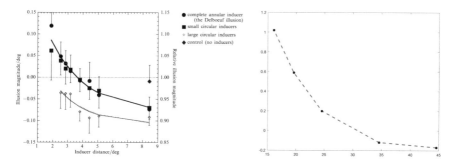

Fig. 5.16 Left: experimental analysis of the decay of the illusion magnitude as a function of the distance between the target and the inducers, from Roberts et al. (2005). The circles refers to the Delboeuf illusion. Right: the graph shows how the perceived displacement decreases as a function of the distance in our simulations. In the abscissa we put the distance $|x' - x|$ and in the ordinate the computed displacement. Our results are in agreement with the ones shown in this experiment

produced by inducers and distance. Target size was fixed at $1.27°$ (visual angle). The reported thickness for the annulus is of $0.63°$, and it was presented at distances of $1.90, 2.53, 2.85, 3.17, 3.80, 4.44, 5.07,$ or $8.45°$ (visual angle). Four subjects were asked to evaluate the illusion magnitude, and the experiment was repeated twice. The results are presented in Fig. 5.16, (left): the black dots refer to the Delboeuf illusion.

We repeated an analogous simulation applying our model. In Fig. 5.15 (first row) we considered five stimuli with a decreasing size of the annulus $\rho(x')$. In the second row we display the computed displacement through the strain theory approach introduced in Sect. 5.3.5. Finally, the third row contains the reconstructed central target, through formula (5.14). The red circle is the target reference of the initial stimulus, drawn in order to allow a comparison. The distance between the center of the annulus and the center of the target $|x' - x|$ decreases from left to right and represents a quantity strictly related to the sizes of the annulus $\rho(x')$. In fact, increasing ρ_1 means we increase the distance between the target and the circumference. However the magnitude of the illusion does not depend on the distance alone: when the annulus is big, we perceive a shrinking (see page 454 of Girgus et al. (1972)), while if the annulus size is decreased, we observe an enlargement of the central target. This variation is explained by an evaluation of the difference in size between the target and the annulus.

The results we obtained are collected in Fig. 5.16 (left). In the x-axis the distance $|x' - x|$ is reported and along the y-axis the computed displacement. The computed displacement decreases as a function of the distance in our computations. The results are in good agreement with the experimental findings of Roberts et al. (2005), represented in Fig. 5.16 (left).

5.5 Conclusions

We proposed a quantitative model for scale/size-GOIs, inspired by the geometry of the visual cortex, (Sarti et al. 2009) and by a model of orientation-related illusions (Franceschiello et al. 2017b). We provided here a very general formulation of the neurogeometrical deformation model in terms of retinal position (x, y) and by a feature f. This allows to choose the scale as encoded feature, and consequently to adapt the model to scale/size-GOIs. The model is then validated onto the Ebbinghaus and Delboeuf illusions, and further compared with the results contained in Roberts et al. (2005), and Massaro and Anderson (1971). The Ebbinghaus and Delboeuf illusions have been often studied together, but it was not clear how to correctly identify the interplay of features playing a role in the second one (Koffka 2013, Gibson 1960). The role played by the distance between the inducers and the target was well established, but this was not sufficient to explain the phenomenon. In our model, we conjectured that the misperception is induced by the size difference between the target and the annulus. This is compatible with existing experiments, and correctly explains the perceptual effect induced by the phenomena. The advantage of a general formulation of the model consists into the possibility to extend the current same idea to other features. We think that it would be possible to extend the current formulation to other GOIs, as for example illusions of movement or those involving higher dimensions encoding.

References

Amari, S. (1972). Characteristics of random nets of analog neuron-like elements. *IEEE Transactions on Systems, Man, and Cybernetics*, pp. 1–24

August, J., & Zucker, S. W. (2000). The curve indicator random field: Curve organization via edge correlation. In *Perceptual organization for artificial vision systems* (pp. 265–288). Springer

Bremner, A. J., Doherty, M. J., Caparos, S., De Fockert, J., Linnell, K. J., & Davidoff, J. (2016). Effects of culture and the urban environment on the development of the e bbinghaus illusion. *Child Development*, 87(3), 962–981.

Citti, G., & Sarti, A. (2006). A cortical based model of perceptual completion in the roto-translation space. *Journal of Mathematical Imaging and Vision*, 24(3), 307–326.

Daugman, J. G. (1985). Uncertainty relation for resolution in space, spatial frequency, and orientation optimized by two-dimensional visual cortical filters. *JOSA A*, 2(7), 1160–1169.

DeAngelis, G. C., Ohzawa, I., & Freeman, R. D. (1995). Receptive-field dynamics in the central visual pathways. *Trends in Neurosciences*, 18(10), 451–458.

Delboeuf, F. J. (1865). Note sur certaines illusions d'optique: Essai d'une théorie psychophysique de la maniere dont l'oeil apprécie les distances et les angles. *Bulletins de l'Académie Royale des Sciences, Lettres et Beaux-arts de Belgique*, 19, 195–216.

Doherty, M. J., Tsuji, H., & Phillips, W. A. (2008). The context sensitivity of visual size perception varies across cultures. *Perception*, 37(9), 1426–1433.

Duits, R., & Franken, E. (2010a). Left-invariant parabolic evolutions on se (2) and contour enhancement via invertible orientation scores part i: Linear left-invariant diffusion equations on se (2). *Quarterly of Applied Mathematics*, 68, 255–292.

Duits, R., & Franken, E. (2010b). Left-invariant parabolic evolutions on se (2) and contour enhancement via invertible orientation scores part ii: Nonlinear left-invariant diffusions on invertible orientation scores. *Quarterly of Applied Mathematics*, 68, 293–331.

Eagleman, D. M. (2001). Visual illusions and neurobiology. *Nature Reviews Neuroscience*, 2(12), 920–926.

Ehm, W., & Wackermann, J. (2012). Modeling geometric-optical illusions: A variational approach. *Journal of Mathematical Psychology*, 56(6), 404–416.

Fermüller, C., & Malm, H. (2004). Uncertainty in visual processes predicts geometrical optical illusions. *Vision Research*, 44(7), 727–749.

Fonteneau, E., Goldstein, J., & Davidoff, J. (2008). Cultural differences in perception: Observations from a remote culture. *Journal of Cognition and Culture*, 8(3–4), 189–209.

Franceschiello, B., Sarti, A., & Citti, G. (2017a) Mathematical models of visual perception for the analysis of geometrical optical illusions. springer INdAM series book. In *Mathematical and theoretical neuroscience: Cell, network and data analysis*. https://doi.org/10.1007/978-3-319-68297-6_9

Franceschiello, B., Sarti, A., & Citti, G. (2017b). A neuro-mathematical model for geometrical optical illusions. *Journal of Mathematical Imaging and Vision*, 60(1), 94–108. https://doi.org/10.1007/s10851-017-0740-6.

Ermentrout, G. B., & Cowan, J. D. (1979). Temporal oscillations in neuronal nets. *Journal of Mathematical Biology*, 7(3), 265–280.

Geisler, W. S., & Kersten, D. (2002). Illusions, perception and bayes. *Nature Neuroscience*, 5(6), 508–510.

Gibson, J. J. (1960). The concept of the stimulus in psychology. *American Psychologist*, 15(11), 694.

Girgus, J. S., Coren, S., & Agdern, M. (1972). The interrelationship between the ebbinghaus and delboeuf illusions. *Journal of Experimental Psychology*, 95(2), 453.

von Helmholtz, H., & Southall, J. P. C. (2005). *Treatise on physiological optics* (Vol. 3). Courier Corporation

Hering, H. E. (1861). Beiträge zur physiologie. 1-5. Leipzig, W. Engelmann

Hoffman, W. C. (1971). Visual illusions of angle as an application of lie transformation groups. *Siam Review*, *13*(2), 169–184

Hoffman, W. C. (1989). The visual cortex is a contact bundle. *Applied Mathematics and Computation*, *32*(2), 137–167.

Wilson, H. R., & Cowan, J. D. (1972). Excitatory and inhibitory interactions in localized populations of model neurons. *Biophysical Journal*, *12*(1), 1–24.

Hubel, D. H., & Wiesel, T. N. (1977). Ferrier lecture: Functional architecture of macaque monkey visual cortex. *Proceedings of the Royal Society of London B: Biological Sciences*, *198*(1130), 1–59.

Jones, J. P., & Palmer, L. A. (1987). An evaluation of the two-dimensional gabor filter model of simple receptive fields in cat striate cortex. *Journal of Neurophysiology*, *58*(6), 1233–1258.

Knill, D. C., & Richards, W. (1996). *Perception as Bayesian inference*. Cambridge: Cambridge University Press.

Koenderink, J. J., & van Doorn, A. J. (1987). Representation of local geometry in the visual system. *Biological Cybernetics*, *55*(6), 367–375.

Koffka, K. (2013). *Principles of Gestalt psychology* (Vol. 44). Routledge.

Künnapas, T. M. (1955). Influence of frame size on apparent length of a line. *Journal of Experimental Psychology*, *50*(3), 168.

Lubliner, J. (2008). *Plasticity theory*. Courier Corporation

Marsden, J. E., & Hughes, T. J. (1994). *Mathematical foundations of elasticity*. Courier Corporation

Massaro, D. W., & Anderson, N. H. (1971). Judgmental model of the ebbinghaus illusion. *Journal of Experimental Psychology*, *89*(1), 147.

Mumford, D. (1994). Elastica and computer vision. *Algebraic geometry and its applications* (pp. 491–506). Berlin: Springer.

Murray, M. M., & Herrmann, C. S. (2013). Illusory contours: A window onto the neurophysiology of constructing perception. *Trends in Cognitive Sciences*, *17*(9), 471–481.

Murray, M. M., Wylie, G. R., Higgins, B. A., Javitt, D. C., Schroeder, C. E., & Foxe, J. J. (2002). The spatiotemporal dynamics of illusory contour processing: Combined high-density electrical mapping, source analysis, and functional magnetic resonance imaging. *The Journal of Neuroscience*, *22*(12), 5055–5073.

Ninio, J. (2014). Geometrical illusions are not always where you think they are: A review of some classical and less classical illusions, and ways to describe them. *Frontiers in Human Neuroscience*, *8*, 856.

Oppel, J. J. (1855). Uber geometrisch-optische tauschungen. Jahresbericht des physikalischen Vereins zu Frankfurt am Main

Bressloff, P. C., & Cowan, J. D. (2003). The functional geometry of local and horizontal connections in a model of V1. *Journal of Physiology-Paris*, *1*(97), 221–36.

Petitot, J. (2008). Neurogéométrie de la vision. Editions Ecole Polytechnique

Petitot, J., & Tondut, Y. (1999). Vers une neurogéométrie. fibrations corticales, structures de contact et contours subjectifs modaux. *Mathématiques informatique et sciences humaines*, *145*, 5–102.

Petitot, J. E., Varela, F. J., Pachoud, B. E., & Roy, J. M. E. (1999). *Naturalizing phenomenology: Issues in contemporary phenomenology and cognitive science*. Stanford University Press

Roberts, B., Harris, M. G., & Yates, T. A. (2005). The roles of inducer size and distance in the ebbinghaus illusion (titchener circles). *Perception*, *34*(7), 847–856.

Sarti, A., & Citti, G. (2015). The constitution of visual perceptual units in the functional architecture of v1. *Journal of Computational Neuroscience*, *38*(2), 285–300.

Sarti, A., Citti, G., & Petitot, J. (2008). The symplectic structure of the primary visual cortex. *Biological Cybernetics*, *98*(1), 33–48.

Sarti, A., Citti, G., & Petitot, J. (2009). Functional geometry of the horizontal connectivity in the primary visual cortex. *Journal of Physiology-Paris*, *103*(1), 37–45.

Yen, S. C., & Finkel, L. H. (1998). Extraction of perceptually salient contours by striate cortical networks. *Vision Research*, *38*(5), 719–741.

Smith, D. A. (1978). A descriptive model for perception of optical illusions. *Journal of Mathematical Psychology, 17*(1), 64–85.

Von Der Heyclt, R., Peterhans, E., & Baurngartner, G. (1984). Illusory contours and cortical neuron responses. *Science, 224*, 1260–1262.

Walker, E. H. (1973). A mathematical theory of optical illusions and figural aftereffects. *Perception & Psychophysics, 13*(3), 467–486.

Weiss, Y., Simoncelli, E. P., & Adelson, E. H. (2002). Motion illusions as optimal percepts. *Nature Neuroscience, 5*(6), 598–604.

Westheimer, G. (2008). Illusions in the spatial sense of the eye: geometrical-optical illusions and the neural representation of space. *Vision Research, 48*(20), 2128–2142.

Williams, L. R., & Jacobs, D. W. (1997). Stochastic completion fields: A neural model of illusory contour shape and salience. *Neural Computation, 9*(4), 837–858.

Yan, T., Wang, B., Yan, Y., Geng, Y., Yamasita, Y., Wu, J., et al. (2014). Attention influence response of ebbinghaus illusion in the human visual area. *International Information Institute (Tokyo). Information, 17*(1), 335.

Part II
The Arts: Fine Arts

Chapter 6
Neuroscience for an Artist; a Beginning

Emilio Bizzi and Robert Ajemian

Abstract In this essay we describe some of the brain processes that are engaged when artists produce drawings and paintings. Two brain structures have been recently identified as key locations where the abstract neural signals representing the intention to move are transformed into patterns of neural activity suitable for activating the muscles. The first of these locations is the putamen, a section of the basal ganglia, the second is in the spinal cord. It is interesting that different types of modules and combinatorial activities characterize both structures. Finally, we discuss the ability of experts to generalize their learning across radically different contexts.

At the June 2018, meeting "Space-time geometries in the Brain and movement in the Arts' held in Paris, Dr. Bizzi described the processes in the brain that are engaged when artists produce drawings and paintings. The drawing shown in Fig. 6.1, by the highly regarded Andrea del Sarto, a Florentine artist active in the mid 1500's, is an example of a piece of art work combining exceptional technical skills and high artistic value. The drawing shown in Fig. 6.1, is a preparatory study representing the head of St. John the Baptist.

The purpose of the study was to facilitate the making of a full body painting of the saint which is now in Florence (at Palazzo Pitti).

Del Sarto, like all artists of that time in Florence, went through a rigorous and protracted training in the art/skill of design. Giorgio Vasari, a contemporary of Del Sarto who chronicled the life and work of Italian painters from medieval times to the late renaissance, emphatically stressed that the practice of drawing was the foundation of the visual arts; painters achieved perfection by careful, protracted practice with preparatory sketches. As a consequence of this practice, an abundance of drawings were produced, some of which have happily survived.

E. Bizzi (✉) · R. Ajemian
Massachusetts Institute of Technology, McGovern Institute for Brain Research,
Cambridge, MA, USA
e-mail: ebizzi@mit.edu

© Springer Nature Switzerland AG 2021   117
T. Flash and A. Berthoz (eds.), *Space-Time Geometries for Motion and Perception in the Brain and the Arts*, Lecture Notes in Morphogenesis,
https://doi.org/10.1007/978-3-030-57227-3_6

Fig. 6.1 Andrea Del Sarto -
preparatory study
representing the head of St.
John the Baptist

On the basis of Vasari's prescriptions, coupled with what we know about the functional organization and plasticity of the nervous system, we can now be confident that intense sensory-motor practice must result in changes of the cognitive state of painters' central nervous system. In this essay, we will describe the cortical and subcortical changes that result from artists' intense practice. It is the ambitious goal of this essay to show how these changes might facilitate the expression of the artist's ideas.

Figure 6.1, which is a drawing made during Del Sarto's mature years, shows how the skills he acquired through years of practice made it possible to express the complex, conflicting feelings represented on the Baptist's face. The pensive, slightly sad expression of the saint's face reveals his awareness of being a precursor of Christ and a martyr. Del Sarto used his skill to convey something that touches our emotions; he captured the saint in a moment of foreshadowing the glory of his life as one close to God's son and the inevitable doom of an impending cruel death.

In the last 100 years neuro-scientists and clinicians have focused on understanding voluntary movements and discovered that the neural processes that subserve even the simplest everyday actions are incredibly complex and only partially understood. In this essay, we will not try to summarize the large amount of data related to movements' production, but will focus instead on a few key discoveries of the last 20–30 years that have made it possible to gain a deeper, albeit still incomplete, understanding of how an evanescent wish to move is translated into actions (Brincat et al. 2018).

6.1 Preliminary Preparatory Neural Events

Let us now imagine that we could peek into Del Sarto's brain as he begins to draw the face of the handsome, but here inconsequential boy of Fig. 6.1. We would then witness the simultaneous appearance of two sets of neural events: sensory visual activity in the occipital cortical areas and motor related activity in the cortical/subcortical areas.

On the sensory side, as the artist's eyes scan the model's features, retinal cells become active and transmit signals to the visual cortical areas. These signals flow in a rostro-frontal direction and they make contact with the neurons of the middle temporal cortical areas, the posterior inferior temporal and the lateral prefrontal cortex (Heyworth and Squire 2019). What happens at these areas is a gradual transformation of the sensory signals into sequences of "categorical abstractions" that precisely reflect the artist's depiction of what he is seeing. This activity eventually becomes the input to the motor system.

On the motor side, as the artist persists in drawing the model over and over in search of perfection, the relevant neural circuitry is reshaped to embody sensory-motor learning or "traces". These traces are gradually established in all the major components of the motor circuitry, like the premotor and prefrontal cortices as well as the medial temporal cortex.

6.2 The Formation of Motor Memories

From a neural cell perspective sensory-motor learning involves changes in synaptic connectivity—and indeed new synapses are formed between cells that are activated at the same time; basically, memories are established through this correlative process. Described in this way memory formation seems a straightforward process, but, as often happens in neuroscience new peculiar features have emerged which indicate an unexpected degree of complexity, and most importantly, the need to rethink some basic assumptions about how memories are formed. The surprising, paradoxical problem is that synapses have a short life—because they are proteins they are churned over every few days—while, of course, memories remain throughout life.

Recent modelling work has provided an alternative explanation of this quandary. Ajemian et al. (2013) developed a neural network characterized by synapses that are constantly changing even during learning a variety of skills (Ajemian et al. 2013). This type of network would be highly redundant with each neuron contacted by a large number of synapses originating from the circuit's other neurons. Ajemian et al. (2013) showed that in this type of neural circuit many different synapses can give rise to the same input- output processing and this network can perform the same function even if its synapses undergo changes. Thus, memory is not specified by a fixed pattern of synapses, but by patterns of input—output processing. In short, the artist's brain keeps changing, but the artistic expertise remains.

6.3   Basal Ganglia Modularity: The Transformation of Cortical Preparative Activity

Del Sarto in his drawings often made use of models that provided an external representation. Of course, many artists do not always utilize external models; they paint or draw by relying on internal representations. An example of a painting whose representation is not dependent on external models is the famous Autumn Rhythm by Pollock (Jiang et al. 2018). It is certainly interesting that Pollock began by drawing a grid on the canvas lying on the floor. In this way he created a structure that constrained the dripping paint. The resulting painting has rhythm and dynamic energy; the seemingly chaotic contorted lines create a tension that makes this painting a masterpiece of abstract expressionism (Fig. 6.2).

As Del Sarto and Pollock begin drawing and painting they would summon to the forefront their artistic expertise, an action which begins a mode of processing across their landscape of cortical activity. These landscapes represent intention—related preparatory activity, that is, neural activity that acts like a command for the motor system. One of the key areas where preparatory activity is formed is the frontal cortex which is a point of convergence for neural signals emanating from different sensory areas as well as memory-related and attention-related cortical areas (Churchland et al. 2010; Guo et al. 2017; Inagaki et al. 2018; Waskom and Wagner 2017). From

Fig. 6.2   Jackson Pollock-Autumn rhythm

the prefrontal cortex the "preparatory" activity reaches a key subcortical structure: the putamen—a subcortical region that is part of the basal ganglia (Graybiel 1998) (Fig. 6.3).

Two important papers have recently contributed to our understanding of the process that are taking place in the cortico-striatal loop. Markowitz et al. (2018) and Wiltschko et al. (2015) used a confluence of machine learning techniques to detect a finite library of sub-movements represented in basal ganglia of freely moving mice. These sub-movements embody recurring behavioral modules or motifs that exist at the sub-second time scale (350 ms), a scale sufficient to act as building blocks for volitional movements that occur on the scale of seconds. In essence, one can think of these modules as kind of micro pattern generators representing recognizable action segments. As a consequence of the arrival of cortical signals at the putamen, the sub-movements are connected to each other in a non-uniform way with each sub-movement preferentially linked to some modules and not others, depending on existing behavioral constraints. This process of concatenation leads

Fig. 6.3 Schematic—central nervous system

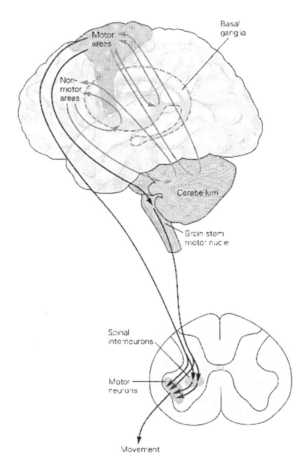

to the formation of full movements and movement sequences. When new behaviors are expressed, new combinations of modules are observed simply by reusing of the same modules without forming any new ones. It is then tempting to conclude that this loop is involved in beginning to implement the sensorimotor transformation by converting abstract movement commands embodied by prefrontal cortical signals into a sequence of movement segments with some "motor" character.

The output of the putamen connects with the cells of the Globus pallidus via D1R and D2R fibers, which connect respectively to the internal/external segments of the Globus Pallidus (Sheng et al. 2019). The Globus pallidus, in turn, connects to the ventro-lateral thalamus and, finally, from there the loop ends on the motor cortex. While the steps from the putamen to motor cortex via globus pallidus and thalamus are meant to make the signals in cortico-striatal loop progressively more "motor like", it is not yet clear whether and how "motor tuning" processes occur at these stations.

The discovery of the putamen's functional properties is certainly an important step; it demonstrates the presence of two neural circuits: a special type of modularity as well as the combinatorial propensity of the micro pattern generator. However, to gain further insight into this region and to firm up what we can consider an important, but initial beginning, much more information about the loop and the putamen should be gathered. The following questions are just an example of what we need to know in order to get an in depth view on the loop and putamen.

1. What kinds of neural patterns are conveyed by the fibers that connect the prefrontal area to the putamen modules? Do these fibers specify a "code" or does the specification emerge after contacting the putamen modules?
2. Does each prefrontal fiber carry a slightly different signal? And if it is so, are the different patterns a variety of motor solutions to be shaped by feedback at some point along the way to the spinal cord?
3. Are the preparatory signals carrying a continuous stream of neural activity from the beginning of the movement to its end or a sequence of segments? (Graybiel 1998; Kadmon et al. 2018).
4. If so, from where are the segments originating? (Abeles et al. 1995; Wymbs et al. 2012).

6.4   Spinal Cord Modularity: The Key Role of Muscle Synergies

To sum, we are fairly confident that the neural signals representing the intention to move are transformed by a loop which begins at the prefrontal cortex and ends at the motor cortex. The end point of the loop are the neurons of the descending cortico spinal tracts; these fibers convey motor signals to spinal motorneurons and interneurons, which ultimately lead to the muscles of the artist hand, arm and body (Dum and Strick 1991).

As mentioned before modularity and combinatorial activity of micro patterns generators are the main functional feature of the putamen. The surprising outcome of recent spinal cord investigations is that modularity and combinatorial activities have also been found to be a spinal cord functional feature (Bizzi et al. 2008; Caggiano et al. 2016).

The neural origin of spinal cord modules rests on studies in several vertebrate species. These studies have demonstrated that a spinal module is a functional unit of spinal interneurons that generates a specific motor output by imposing a specific pattern of muscle activation (muscle synergies). Muscle synergies are neural coordinative structures that function to alleviate the computational burden associated with the control of movement and posture. Anatomically, the modules are made up of groups of spinal interneurons whose efferent fibers make contact with a distinct set of motor neurons (Caggiano et al. 2016; Fetz et al. 2002). It follows that whenever these motor neurons are activated by descending cortico-spinal impulses and/or reflex pathways from the periphery a distinct muscle synergy becomes active. This process leads to the formation of muscle "synergies" which represent kind of functional building blocks whose combination leads to the "construction" of voluntary movements (Bizzi et al.2008; Caggiano et al. 2016). A factorization algorithm that takes as input all the recorded muscle EMG data is utilized to extract muscle synergies and activation coefficients. The factorization procedure essentially performs a dimensionality reduction by grouping muscles that tend to co-vary in the data set into individual synergies.

In the last few years many investigators have examined motor behaviors in humans and animals. The results show that combining a small set of muscle synergies appears to be a general strategy that the central nervous system utilizes for simplifying the control of movements (Bizzi et al. 2008).

All in all the following important points have emerged from spinal cord investigations: (1) the same synergy may be utilized in different motor behaviors, and (2) different behaviors may be constructed by linearly combining the same synergies with different timing and scaling factors, and (3) the development of new skills over long periods of time leads to the formation of new specialized task synergies.

With respect to the question of cortical control of muscle synergies the study of Overduin et al. (2015) demonstrated that intra cortical stimulation in monkeys elicited EMG patterns that could be decomposed into muscle synergies. Importantly, these EMG patterns were found in a few cases to be similar to those evoked during the same animal's voluntary movements. Whether this finding indicates that the cortex "encodes" muscle synergies remains to be determined.

At this point our knowledge of the supraspinal machinery involved in synergy activation is inadequate. We need to know how the signals elaborated by the supraspinal loop play into the spinal modules.

On the bright side, the experimental evidence summarized above indicates that the spinal cord operates as a discrete combinatorial system. In a way, the motor system is like language, a system in which a discrete elements and a set of rules for combining them can generate a large number of meaningful entities that are distinct from those of their elements. And just as with language, the combinatorial system can accommodate different levels of expertise.

6.5   Generalization

While we can reasonably assume that progress will eventually be made on some of the questions discussed in this essay, there are certain aspects of motor behaviors that seem harder to understand. Motor generalization is a case in point and Michelangelo Figs. 6.4 and 6.5 illustrate an example of this behavior.

Figure 6.4 is a preparatory study of a nude male which Michelangelo utilized for the painting of a section of the ceiling of the Sistine Chapel. Note that the male represented in Fig. 6.5 was painted while Michelangelo was hoisted parallel to the chapel's ceiling. Because the postural changes lead to changes in the influence of gravity on Michelangelo's brush strokes, the motor output needed to implement the same artistic vision has changed entirely! Yet Michelangelo, as an expert, effortlessly

Fig. 6.5 Michelangelo—Ignudo—ceiling of Sistine Chapel

Fig. 6.4 Michelangelo—
seated nude and two studies
of an arm, about 1510-12

accomplishes the task. This is known as generalization, because learning in one set of circumstances applies more generally to other circumstances.

   To paint in that awkward and uncomfortable posture requires, at the minimum, a radical re-programming of most of Michelangelo's muscles. Whether the reprogramming was initiated by a massive change of proprioceptive feedback is no more than an educated guess. But, for neuroscientists interested in pursuing complex and certainly challenging questions of how experts generalize, this motor behavior might be an opportunity to explore neural complexity through creative computer modelling.

   Broad generalization embodies the highest form of expertise in any sort of skill learning—and one which as of yet still eludes current approaches in Artifical Intelligence. Michelangelo's extraordinary artistic ability involves multiple forms of generalization, many of which involve his ability to utilize highly sophisticated mental models of the geometry of both space and time. Here we categorize multiple types of mappings across which Michelangelo generalized to produce his art.

   Artistic concept to 2-D image: Michelangelo invariably has a 3-D model in his head of body shape and form. Yet the rendering takes place on a 2-D surface. In order to make this transformation from a 3-D idea to a 2-D form, Michelangelo possesses complete mastery of projective geometry and linear perspective, as formally articulated by Filippo Brunelleschi at the onset of the Renaissance.

Temporal motion to curvilinear static form: More so than in the case of most works of art, the frescoes of the Sistine Chapel are bursting with a latent dynamic energy of motion, perhaps most famously illustrated by the finger of God reaching out to touch Adam in the *Creation of Adam*. Here, Michelangelo is able to masterfully convey the dynamics of human movement through curvilinear static forms.

Motor intentions to motor output in a completely novel posture: Michelangelo painted the Sistine Chapel while standing on a scaffold with his head uncomfortably tilted back and his arms reaching upwards. To paint in that awkward and uncomfortable posture requires, at the minimum, a radical re-programming of most of Michelangelo's muscles. Part of this reprogramming may have been initiated by a massive change of proprioceptive feedback to provide feedback guidance for his brushstrokes. Still, even these feedback signals would be of an unfamiliar form; yet without any specific practice of painting in this posture and even with the acute time constraints of having to finish each fresco while it is wet, Michelangelo nonetheless managed without requiring any "do-overs".

These are just some of the many mappings which were integrated by Michelangelo and across which he generalized adeptly in order produce a masterful work of art. For neuroscientists interested in pursuing complex and certainly challenging questions of how experts generalize, the skills of an artist might be an opportunity to explore neural complexity underlying sophisticated sensorimotor control.

References

Abeles, M., Bergman, H., Gat, I., Meilijson, I., Seidemann, E., Tishby, N., & Vaadia, E. (1995). Cortical activity flips among quasi-stationary states. *Proceedings of the National Academy of Sciences, 92*(19), 8616–8620.

Ajemian, R., D'Ausilio, A., Moorman, H., & Bizzi, E. (2013). A theory for how sensorimotor skills are learned and retained in noisy and nonstationary neural circuits. *Proceedings of the National Academy of Sciences, 110*(52), E5078–E5087.

Bizzi, E., Cheung, V. C. K., d'Avella, A., Saltiel, P., & Tresch, M. (2008). Combining modules for movement. *Brain Research Reviews, 57*(1), 125–133.

Brincat, S. L., Siegel, M., von Nicolai, C., & Miller, E. K. (2018). Gradual progression from sensory to task-related processing in cerebral cortex. *Proceedings of the National Academy of Sciences, 115*(30), E7202–E7211.

Caggiano, V., Cheung, V. C., & Bizzi, E. (2016). An optogenetic demonstration of motor modularity in the mammalian spinal cord. *Scientific Reports, 6,* 35185.

Churchland, M. M., Cunningham, J. P., Kaufman, M. T., Ryu, S. I., & Shenoy, K. V. (2010). Cortical preparatory activity: Representation of movement or first cog in a dynamical machine? *Neuron, 68*(3), 387–400.

Dum, R. P., & Strick, P. L. (1991). The origin of corticospinal projections from the premotor areas in the frontal lobe. *Journal of Neuroscience, 11*(3), 667–689.

Fetz, E. E., Perlmutter, S. I., Prut, Y., Seki, K., & Votaw, S. (2002). Roles of primate spinal interneurons in preparation and execution of voluntary hand movement. *Brain Research Reviews, 40*(1–3), 53–65.

Graybiel, A. M. (1998). The basal ganglia and chunking of action repertoires. *Neurobiology of Learning and Memory, 70*(1–2), 119–136.

Guo, Z. V., Inagaki, H. K., Daie, K., Druckmann, S., Gerfen, C. R., & Svoboda, K. (2017). Maintenance of persistent activity in a frontal thalamocortical loop. *Nature, 545*(7653), 181.

Kadmon Harpaz, N., Ungarish, D., Hatsopoulos, N. G., & Flash, T. (2018). Movement decomposition in the primary motor cortex. *Cerebral Cortex, 29*(4), 1619–1633.

Heyworth, N. C., & Squire, L. R. (2019). The nature of recollection across months and years and after medial temporal lobe damage. *Proceedings of the National Academy of Sciences, 116*(10), 4619–4624.

Inagaki, H. K., Inagaki, M., Romani, S., & Svoboda, K. (2018). Low-dimensional and monotonic preparatory activity in mouse anterior lateral motor cortex. *Journal of Neuroscience, 38*(17), 4163–4185.

Jiang, J., Wagner, A. D., & Egner, T. (2018). Integrated externally and internally generated task predictions jointly guide cognitive control in prefrontal cortex. *Elife, 7,* e39497.

Markowitz, J. E., Gillis, W. F., Beron, C. C., Neufeld, S. Q., Robertson, K., Bhagat, N. D., et al. (2018). The striatum organizes 3D behavior via moment-to-moment action selection. *Cell, 174*(1), 44–58.

Overduin, S. A., d'Avella, A., Roh, J., Carmena, J. M., & Bizzi, E. (2015). Representation of muscle synergies in the primate brain. *Journal of Neuroscience, 35*(37), 12615–12624

Sheng, M. J., Lu, D., Shen, Z. M., & Poo, M. M. (2019). Emergence of stable striatal D1R and D2R neuronal ensembles with distinct firing sequence during motor learning. *Proceedings of the National Academy of Sciences, 116*(22), 11038–11047.

Waskom, M. L., & Wagner, A. D. (2017). Distributed representation of context by intrinsic subnetworks in prefrontal cortex. *Proceedings of the National Academy of Sciences, 114*(8), 2030–2035.

Wiltschko, A. B., Johnson, M. J., Iurilli, G., Peterson, R. E., Katon, J. M., Pashkovski, S. L., et al. (2015). Mapping sub-second structure in mouse behavior. *Neuron, 88*(6), 1121–1135.

Wymbs, N. F., Bassett, D. S., Mucha, P. J., Porter, M. A., & Grafton, S. T. (2012). Differential recruitment of the sensorimotor putamen and frontoparietal cortex during motor chunking in humans. *Neuron, 74*(5), 936–946.

Chapter 7
Art and the Geometry of Visual Space

Alistair Burleigh, Robert Pepperell, and Nicole Ruta

Abstract This chapter considers the geometric structure of visual space and the way it has been artistically depicted. This structure is hard to quantify using scientific methods and has not been precisely defined, in part because it is highly variable and subjective. But artists have often sought to record their visual experience and in doing so have rendered the structure of visual space in the objective form of paintings and drawings. Where audiences respond favourably to these artworks, it is in part because they recognise in them aspects of their own visual experience. Artworks, then, can serve as a source of data from which we derive evidence about the nature of visual space, and this in turn can inform scientific investigation of its geometry. It has been argued that linear perspective—a projective geometry discovered by artists and architects in fifteenth-century Italy—is the most accurate way to depict visual space. But although often trained in the techniques of linear perspective, artists have rarely applied its rules rigorously, and have instead developed various forms of 'natural perspective' that, it can be argued, are more effective at depicting visual space and so operate more effectively as works of art.

'Scientific perspective is nothing but eye fooling illusionism; it is simply a trick – a bad trick – which makes it impossible for an artist to convey a full experience of space, since it forces the objects in a picture to disappear away from the beholder instead of bringing them within his reach.' Georges Braque, in Richardson (1953)

7.1 Introduction

In this chapter we focus on the way artists have depicted visual space, and what their work reveals about both its phenomenology and its geometry. We will see that visual space does indeed have a broadly defined geometrical structure, although a precise mathematical description remains to be worked out. This structure is reflected in

A. Burleigh · R. Pepperell (✉) · N. Ruta
Fovolab, Cardiff Metropolitan University, Cardiff CF5 2YB, UK
e-mail: rpepperell@cardiffmet.ac.uk

© Springer Nature Switzerland AG 2021
T. Flash and A. Berthoz (eds.), *Space-Time Geometries for Motion and Perception in the Brain and the Arts*, Lecture Notes in Morphogenesis,
https://doi.org/10.1007/978-3-030-57227-3_7

the way artists organise their depictions of space and this in turn has consequences for how we perceive those depictions as viewers. The structure can be most clearly appreciated when we compare artistic depictions of visual space to linear perspectival ones. The geometry of linear perspective is well defined and has been proposed as the most accurate way to depict visual space (Pirenne 1970). But as we will see, artists have often avoided using linear perspective in any mathematically consistent way. Instead they have often employed various kinds of 'natural perspective'. Natural perspectives provide, in effect, pictorial records of the phenomenal structure of vision and offer a basis on which to mathematically model visual space geometry. They may also yield important clues about why certain artworks are aesthetically effective, and how imaginary space is structured.

7.2   Linear Perspective and Its Limitations

The most widely propagated story about how art in the European tradition developed is that prior to the Italian Renaissance painters lacked any understanding of optics, projective geometry, or how to make an image look perceptually realistic. They represented objects in space according to a naïve or primitive set of pictorial conventions in which the size of figures, for example, was determined by a symbolic hierarchy rather than optically-based rules of construction (White 1972). Following the discovery of linear perspective in the early fifteenth century, painters abandoned their previous habits and applied a rigorous mathematical structure to their work. This resulted in an increased sense of depth and heightened levels of realism in their images.

   While this story contains elements of fact, it is also simplistic and misleading. Linear perspective is a beautifully elegant, relatively straightforward, and objectively accurate method of mapping light rays from three-dimensional space to a picture plane. But it also has severe limitations that were known even to early adopters such as Leonardo da Vinci and Piero della Francesca (Kemp 1990). These limitations prevented linear perspective from being universally and rigorously applied by artists, and ultimately led to it being largely abandoned. Painters today hardly ever use it.

   One major limitation of linear perspective is that it is impractical for representing wide angles of view in a way that appears naturalistic under normal viewing conditions. Linear perspective pictures require the viewer's eye to be located at the correct centre of projection in front of the picture to achieve the best illusory effect (Todorović 2009). For any given linear perspective image this distance is equivalent to the focal length of the lens needed to project the image (Kingslake 1992). For wide-angle images with short focal lengths (e.g. <25 mm) viewing distances are less than can be comfortably accommodated by the human eye. Enlarging the image can increase accommodation distances but for a very wide angle of view (e.g. >170° horizontally) the amount of enlargement needed would be impractical since the focal length is effectively zero. The result of viewing linear perspective images from a distance greater than the centre of projection is that objects in line with the principal ray can appear unnaturally minimised, while objects in the periphery can appear unnaturally stretched. This is one source of the dissatisfaction expressed by the Cubist painter Georges Braque in the quote above.

There are several other ways in which linear perspective is limited in its capacity to represent human visual experience naturalistically. Most of us see with two eyes and so enjoy the heightened depth sensation that stereoscopic vision can provide, but linear perspectival images are generally monocular. Visual space is curved, due in part to the physical shape of the eyeball and in part to the processing of visual stimuli by the brain (von Helmholtz 1866; Panofsky 1924), but linear perspective, as the name states, is linear. The acuity of the human visual field is non-homogenous, the relatively small area of the central visual field having much higher acuity than the far peripheral field (Anstis 1998), but the geometry of linear perspective does not differentiate between central and peripheral acuity. Over the course of post-Renaissance European art history, the inability of linear perspective to capture such features of vision seems to have discouraged artists from applying its rules in any consistent or systematic way and led instead to the emergence of various 'natural' perspectives.

7.3 Natural Perspectives and Their Controversies

Being unable to fully rely on accurate linear perspective, artists evolved alternative methods of depicting visual space more suited to their professional needs. These can be broadly categorised as 'natural perspectives' as distinct from 'artificial perspective' based on the geometry of optics. It was Leonardo da Vinci who distinguished between natural and artificial perspective—between the way things appear in natural vision and how they are rendered by linear projective geometry on a plane—and according to John White he conceived a curvilinear method of rendering three-dimensional space that ameliorated some of the inherent problems of linear perspective (White 1972). Many further nonlinear natural or subjective perspective systems were subsequently developed or theorised, including by Parsey (1840), Herdman (1853), Hauck (1879), Walters (1940), Hansen (1973), Rauschenbach (1982), and Floçon and Barre (1988).

These systems are sometimes derived from empirical observation and practical experimentation and sometimes from geometric principles. They are concerned with depicting the way we look at the world and the way we look at depictions of what artists see when they look at the world. The features of visual perception typically addressed by these natural perspective systems are those that are especially problematic for linear perspective, including wide fields of view (binocular human vision extends some 180° horizontally), the effects of binocularity, the apparent curvature of objectively straight lines (which is especially noticeable in the peripheral visual field), variations in acuity across the visual field, and the apparent enlargement of objects occupying the central focus of the visual field (Pepperell and Hughes 2015).

Natural perspective systems have proved controversial throughout their history, and to some extent remain so today. They are open to the criticism that by emulating features of subjective perception the artist is unnecessarily duplicating or exaggerating perceptual effects that would occur anyway when an optically accurate image is viewed (Pirenne 1970; Gombrich 1982; Tyler 2015). For example, enlarging an object in the picture in order to give it the same prominence it would have when viewed in natural vision will cause it to suffer a double enlargement when it is viewed again in the picture. Likewise, straight lines rendered as curved, as seen in the peripheral visual field, will undergo additional curvature when seen in the peripheral field of the viewer of the painting. Critics of natural perspective have argued that all these subjective effects 'come for free' if a painting is constructed according to the laws of linear perspective and the viewer's eye is correctly positioned in relation to it.

Defenders of natural perspective systems might reply in two ways. First that such criticism overlooks the practical problems that artists face in determining the location of the viewer with respect to their work. In most cases artists have little control over this, and so the artwork must be aesthetically effective from a wide range of non-ideal viewing positions. Second, with the possible exception of the trompe l'oeil genre, artists are not interested in replicating reality (which is technically near impossible when spanning the full binocular visual field) but in maximising the aesthetic effectiveness of their work. Centuries of artistic experimentation have shown that maximising aesthetic effect can often be best done by emulating, and even exaggerating, certain salient features of visual perception in spite of, or perhaps because of, the apparent duplication of perceptual effects this entails. Consider, for example, the advice given by Du Fresnoy (1667) to seventeenth century painters seeking to achieve greater depth and volume in their work:

> As when we see in a convex mirror, the figures and all other things advancing more strongly and vividly than even natural objects do, and the vivacity of the colours is increased in the parts full in your sight, while the goings off are more and more broken and faint as they approach to the extremities; in the same manner may bodies be given relief and roundness.

The convex mirror he probably had in mind was a device made of polished metal or mineral, such as a Claude Glass, commonly used by artists in the seventeenth century to reflect scenes from nature they wished to paint (Maillet 2009). The image reflected in such a mirror displays some of the subjective visual features noted above, such as the relatively enlarged central area and diminished and somewhat curved periphery compared to a flat mirror. By replicating these features in their pictures, artists are, according to Du Fresnoy, outdoing nature itself in pursuit of stronger and more vivid aesthetic effect. But this approach was thought to be at odds with artistic ideals informed the then dominant theory of artificial or linear perspective. Around the same time Du Fresnoy was advocating a version of natural perspective, Bosse (1649) was decrying artists who relied on it:

> They do not understand how to make objects recede and turn into the distance by using an arrangement of parallel planes. For these gentlemen are, as you know, accustomed to making an entire painting have the same spherical, mirroring effect, which amounts to representing objects the same way the eye sees them, which is an absolutely false and ridiculous thing to propose.

This tension between artificial and natural perspectives—between methods aimed at depicting objective patterns of light and those designed to depict how that light is subjectively experienced—remained at the core of art theory for several centuries (Panofsky 1924; Kemp 1990; Elkins 1994). It surfaced, for example, in debates about whether or not artists should include stereoscopic data in their paintings, a tendency that briefly flourished among avant-garde Victorian painters in Britain and of which the critic John Ruskin disapproved (Ruskin 1873). It also surfaced as an issue of some public prominence during the early nineteenth century in Britain when the artist Arthur Parsey proposed a novel method of perspective construction that he claimed more faithfully represented how we see than was possible using conventional linear perspective (Parsey 1840; Kemp 1987). This precipitated a prolonged and heated debate, in which Ruskin also became involved, about whether it was preferable to depict vertically rising forms, such as towers, with or without convergence (Bantjes 2014).

Artists have continued to develop various forms of natural perspective that more faithfully represent the experience of visual space within the limitations of the medium. Some recent examples include Downes (2005), Mann and Mann (2008), Paraskos (2010), Hockney (2015) and Burleigh et al. (2018a). These approaches have in common a rejection of linear perspectival construction and a tendency towards curvilinear or non-linear forms of projection.

7.4 Natural Perspectives in Art and Science

Paul Cézanne was a prodigious landscape painter who worked by transcribing what he saw in nature directly to the canvas and is sometimes thought to have recorded certain subjective features of perception, such as the relative indistinctness of the visual periphery (Pepperell 2012). Analysis of the composition of his paintings shows that they are not constructed according to the laws of linear perspective (Loran 1963; Rauschenbach 1982; Pepperell and Haertel 2014). When comparing linear perspective photographs of landscapes that he painted to his paintings we find that Cézanne consistently enlarged the central motif and compressed the peripheral regions. An example can be seen in Fig. 7.1, where the central motif in the painting—the well and millstone—appears significantly larger in the painting than the photograph, despite both images showing approximately the same physical space. This pattern recurs frequently throughout art history, which suggests artists have long been employing a pictorial structure that corresponds in some way to the geometry of visual space, that is, to how physical space appears in perception. Here we briefly survey a few examples in which attempts by artists, and scientists, to depict visual space can be compared to linear perspective equivalents, and how this might reveal its geometrical structure.

In the early nineteenth century, the British painter John Constable recorded views of landscapes by framing them within a wooden device fitted with a glass plate on which he traced outlines of objects in the scene and then transferred them to paper

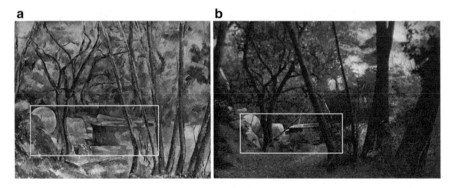

Fig. 7.1 **a** is a reproduction of Paul Cézanne's *Well: Millstone and Cistern Under Trees (Meule et citerne sous bois)*, 1892, oil on canvas, The Barnes Foundation, Merion, Pennsylvania and **b** shows a photograph of the same scene by John Rewald (from Loran 1963). Both pictures are made from the approximately same position and show approximately the same physical space. Note that while the painted version (**a**) is smaller than the photograph, despite showing the same physical space, the central motif of the painting—the well and millstone contained in the white box (added by the authors)—is significantly larger in the painting than in the photograph. *Image copyright* Reproduced with permission from Pepperell and Haertel (2014)

(Kemp 1990; Parris and Fleming-Williams 1991). Figure 7.2a shows an example of a drawing made with this device. According to Alfred Parsey, who witnessed it in use, Constable used four pieces of string tied to each corner of the frame and held at an apex in his mouth to keep his head still and his eye in the correct position (Parsey 1840). The drawing is, in effect, a linear perspective projection of the scene, with the artist's eye being located at the centre of projection. When we compare his optically derived projection to the final paintings made using them, such as that shown in Fig. 7.2b it is notable that despite this optically accurate template, Constable deviated markedly from linear perspective in the final composition. The central area, containing the lower halves of the trees to the right and the right hand river bank, align almost perfectly between the drawing and the painting, as can be seen by comparing the a and b in Fig. 7.2. But the areas around this central section are significantly compressed in the painting, notably the leftmost barge, the houses in the distance, and the upper section of the tree on the right. Since the drawing and the painting depict the same slice of visual space, and the painting is smaller than the drawing, Constable has clearly adjusted the composition in a way that gives greater scale and prominence to the central area.

William Herdman was another nineteenth century British painter of the land-scape, now best remembered for the paintings he made of his home city of Liver-pool. But he also earned income depicting major European cityscapes popular with tourists. Herdman found that the linear perspective techniques in which he had been trained were not helpful for recording the impressive wide vistas he encountered during his travels. Driven by the practical need to convey this expansive visual space to his buyers, he developed a novel geometric method of curvilinear construction. This, he claimed, allowed him to capture far greater spans of physical space with a

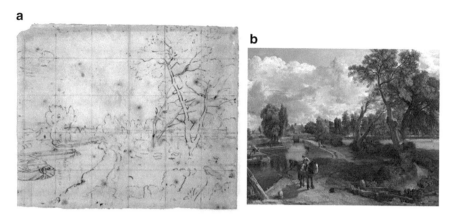

Fig. 7.2 a *Study for Flatford Mill* c. 1816, graphite on paper, Tate Collection, London. **b** *Scene on a Navigable River (Flatford Mill)*, John Constable, oil on canvas, 1816–7, Tate Collection, London. Constable has followed the design in the study closely in the central area of the picture. If the painting is overlaid on the drawing the trunks of the trees in the right half of the painting align almost perfectly, as does the right-hand line of the river bank. However, towards the margins of the scene the painting and the drawing start to deviate markedly. The upper parts of the trees on the right, for example, have been vertically compressed in the painting, while the trees and houses on the horizon to the left have been horizontally compressed. The painting shows the same overall field of view as the drawing but occupies less space and is therefore smaller. Constable seems to have adjusted the scale of the objects in this part of the painting in order to fit more visual space of the scene into the frame without diminishing the central areas. *Image copyright* **a** Tate Gallery, London: Image Released Under Creative Commons; **b** Tate Gallery, London: image released under creative commons

more natural appearance than could be achieved using standard linear perspective (Herdman 1853). While his method later acquired some mathematical rigour, his initial experiments were based on direct, empirical observations in which he tried to transcribe a physical space to the picture plane in a way that corresponded most closely to what he saw. The drawing he made from inside the Rosslyn Chapel in Scotland, reproduced in Fig. 7.3a, shows a wide angle of view, captured relative to a fixation point that is marked with a circle towards the lower left of the drawing bisected by a horizontal line. Herdman recounts: "This drawing was made with the most perfect accuracy according to vision; that is to say every line in it was drawn *exactly* as appeared to the eye from a given spot" (Herdman 1853, emphasis in the original).

A standard linear perspective projection of such a wide angle of view would render the columns to the left and right of the scene as horizontally stretched, and without the subtle curvature in the foreground flagstones. Figure 7.3b shows a contemporary photograph superimposed on Herdman's drawing. The photograph is, in effect, a linear perspective projection taken from the same position in the chapel and directed at the same fixation point used by Herdman. The drawing and the photograph have been aligned as closely as possible so that the architectural features around Herdman's fixation point correspond in scale and position. Note that the column on the left is

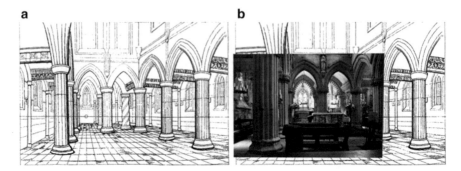

Fig. 7.3 **a** Reproduces a drawing of the interior of Rosslyn Chapel by William G. Herdman, from Herdman (1853). The circle in the arch to the left of centre, bisected by the horizontal line, marks the fixation point relative to which the scene was drawn. **b** Shows a photograph of the interior of taken from the same position from which Herdman made his drawing. The photograph has been superimposed on the drawing such that the architectural features around the fixation point align with those in the drawing. Note the relative size of the leftmost column in the photograph compared to its equivalent in the drawing, the former having undergone a much greater degree of enlargement, both vertically and horizontally, compared to the drawn version. Note also the greater amount of physical space included within the frame of the drawing on the left side compared to the photograph. *Image copyright* **a** public domain; **b** photograph copyright Stuart Jeffreys, 2016. Reproduced with permission

located further to the right in the drawing than in the photograph but is approximately half the width, indicating relative overall spatial compression in this part of the scene. As a result of this spatial compression more of the physical space in the left part of the chapel is visible in the drawing than the photograph.

In 1884 the post-Impressionist painter Georges Seurat made his famous composition *Bathers at Asnières*, now in the National Gallery, London. It depicts, using a relatively wide-angle view, a group of figures relaxing on the banks of the river Seine, and is rendered in Seurat's signature pointillist style, in which he built up passages of colour using small dabs of pigment. Seurat composed the picture carefully, making several small studies, which can also be seen in the National Gallery. One striking aspect of the painting is the way it deviates from the rules of linear perspective, although viewers are unlikely to be consciously aware of it. The central seated figure is far larger than it would be if the scene had been rendered using a standard wide-angle projection. This can be seen in Fig. 7.4, where a modified reproduction of the painting is shown next to the original composition.

In the modified version (b), the size of the figures has been adjusted by one of the authors (using Photoshop) so that they match more closely the relative sizes they would be if rendered using wide-angle linear perspective. The figure lying in the foreground, for example, is larger in the modified version than in the original (a) because it is closer, and the central seated figure is smaller because it is further away. However, it is unlikely Seurat composed the picture as he did out of ignorance of the rules of linear perspective; as a student who attended the nineteenth-century École des Beaux-Arts he would have been rigorously trained in the method. More likely

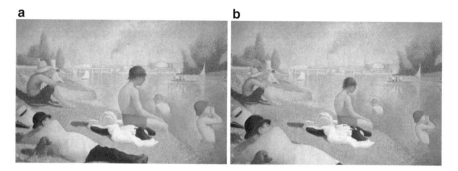

Fig. 7.4 a Georges Seurat, *Bathers at Asnières*, 1884, Oil on canvas, 201 × 300 cm. National Gallery, London. Image source and credit: Wikicommons. **b** A version of *Bathers at Asnières* modified to conform to more closely to a linear perspectival geometry, as generated by one of the authors. Note that in (**b**) the central seated figure is much smaller than in the original version, and the figure lying in the foreground is much larger, as would be the case if the scene were constructed according to the rules of linear perspective. *Image copyright* **a** Wikicommons; **b** created by one of the authors (R Pepperell) by adapting (**a**)

it was composed like this for aesthetic reasons, to emphasise the prominence of the central figure in the space, which is the focal point of the painting. Indeed, when the painting is viewed in situ from close proximity (at roughly the same distance from the surface that the artist was when he painted it), and the focus of attention is given to the head of the central figure, one feels an uncanny sense of space and depth. This form of non-linear natural perspective is effective at conveying the phenomenal experience of visual space.

Throughout the twentieth century we find many artists experimenting with natural perspective methods of depicting visual space. In these cases, a pattern consistent with that outlined above often recurs, namely that regions of space in the centre of the scene tend to be enlarged as compared with linear perspective and those in the periphery tend to be diminished. Examples can be found among the interior scenes that Pierre Bonnard painted in southern France during the early part of the twentieth century (Pepperell 2016). Bonnard often painted unusually wide fields of view and was fascinated with the elusive properties of peripheral vision (Clair 1984). In the 1980s, long after Bonnard's death, the British artist Sargy Mann and a colleague took a series of photographs in Bonnard's house that captured, in linear perspective form, the same views Bonnard had painted (Mann 1994). When we compare Bonnard's paintings to the photographs, we find that the artist often enlarged objects in the central region of the scene and diminished those in the periphery, especially in the region of space closest to the viewer and in the lower part of the visual field. The result is effectively a reversal of the most basic principle of linear perspective, which requires objects to appear larger as they approach the viewer and smaller as they recede. The same principle was employed years later by Sargy Mann when painting his own wide-angle views of scenes (Mann and Mann 2008).

It is not exclusively artists who have attempted to record the experience of visual space in non-linear perspectival ways, for we also should note the work of Blanche Ames who with her brother Adelbert undertook an intensive study of the nature of visual experience and how it can be depicted (Behrens 1987). Adelbert Ames is now better known as an important psychologist of visual perception, but early in his career he had aspirations to be an artist and worked with Blanche on a series of paintings and several related scientific papers. Many of the paintings are concerted attempts to record the effects of binocular 'double vision' in a way similar to the work produced a few decades later by the British artist, Evan Walters (Walters 1940). But in one of the papers they co-authored they attempt to quantify the effect of peripheral compression of visual space and the loss of peripheral acuity noted in the work of many artists above (Ames et al. 1923). Figure 7.5 shows two images from their paper *Vision and the Technique of Art*, illustrating in photographic form the difference between an image as—they claim—it would appear on the retina and as it appears through a 'corrected lens', that is, a linear perspective projection. The 'retinal' version, they argue, is closer to visual appearance in three main respects: first, the central area of the scene is sharper than the peripheral areas, with the result that objects in the periphery are 'softer'; second, the curvature 'barrel distortion' more closely matches the inherent curvature of the retinal image; third, objects in the periphery of the scene appear relatively compressed. The consequence of these effects is an image that the authors claim has more depth, is more natural, and is more pleasing than its rectilinear counterpart.

From this brief survey it is evident that when artists, and occasionally scientists, set out to record their experience of visual space, they often deviate markedly but consistently from linear perspective projection, especially for wide-angled views.

a   b

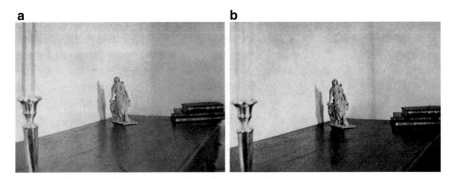

Fig. 7.5 Photographs from Ames et al. (1923) showing what the authors describe as the perceptual distortions of visual space due to the optical and psychological factors. **a** Shows a scene as it appears to the eye, according to the authors, with a degree of barrel distortion, especially evident in the candle on the left, and where slightly more space is visible than in (**b**). Objects in the periphery are in softer focus than in the centre. **b** Shows the 'undistorted' linear perspective projection, which captures less horizontal space and in which all objects are equally sharply focused. *Image copyright* Reproduced from Ames et al. (1923)

This suggests some common, non-linear property of visual space is being empirically recorded in such works. It also suggests, given the art historical reputation of some of the artists mentioned, that the resulting works are aesthetically appealing to audiences. This may help to explain why artists have, by and large, avoided using linear perspective, even though it can in theory provide an optically exact representation of the world, and why audiences admire works of art made using these principles more than they do photographs of equivalent scenes (Hockney 2015). What, then, do these empirically-derived depictions of the visual world reveal about the geometry of visual space?

7.5 The Geometry of Visual Space

The geometrical structure of visual space was first scientifically investigated by Hermann Helmholtz in his *Treatise on Physiological Optics* (von Helmholtz 1866). He used the curved checkerboard pattern in Fig. 7.6a to demonstrate that the visual system does not replicate the rectilinear properties of the Euclidean space we use to describe and measure the physical world. But today there is still very little agreement about exactly what the geometrical structure of visual space is (Wagner 2006; French 2015). Some have proposed that visual space conforms to a hyperbolic curvilinear

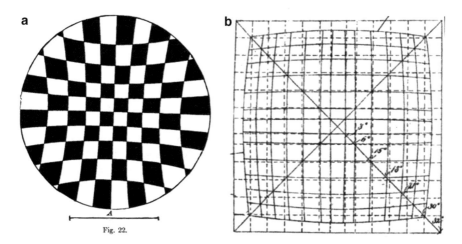

Fig. 7.6 a Reproduces a figure from Hermann von Helmholtz's Treatise on Physiological Optics (von Helmholtz 1866) that demonstrates the inherently nonlinear nature of visual space. In the original image, which is shown larger than here, viewers are asked to look at the centre of the disc with one eye from a distance equivalent to the line below it. From this position the curvature of the checkerboard pattern disappears. **b** Reproduces a figure from Ames et al. (1923) that illustrates the 'barrel distortion' of the retinal image, and the difference between how an image will appear when that distortion is taken into account, as shown by the curvilinear grid overlaid on the rectilinear one

geometry (Luneburg 1947; Heelan 1984). The experimental data on this is inconsistent but does confirm that even if it not entirely hyperbolic visual space is certainly non-Euclidean (Wagner 2006). Indow (2004) claimed our understanding of visual space is 'pluralistic and fragmentary'. Others have argued there is no single geometry for visual space at all (Koenderink and van Doorn 2008; Foley 1972; Suppes 1995; Wagner 2006). In spite of many decades of investigation we are still to "establish a metric for the manifold of visual sensations" (Luneburg 1947).

One reason vision science may have so far failed to define a geometry for the global structure of visual space is that experimental methods tend to measure relatively narrow aspects of the visual field, focusing on either the frontal plane or the depth plane, and on central vision only (Wagner 2006; Yang and Purves 2003). Measuring the perceptual geometry of the full binocular visual field, including the far periphery, presents a number of technical and practical challenges. For example, the standard flat monitors used widely in vision science experiments cover only a relatively small angle of view and being flat do not easily allow for peripheral stimuli to be located equidistantly with the central stimuli from the participant's eyes (Yu and Rosa 2010).

Where studies have been made with wide visual angles or in the far peripheral visual field, the conclusions have been mixed. Different studies reported both the tendency to overestimate (Mateeff and Gourevich 1983; Hubbard and Ruppel 2000; Uddin 2006; Fortenbaugh and Robertson 2011) or underestimate (Bock 1993; Bruno and Morrone 2007; Enright 1995) the eccentricity of static targets in visual space, as well as the tendency to overestimate (Bedell and Johnson 1984) or underestimate (Newsome 1972; Schneider et al. 1978) size of stimuli presented in peripheral visual field. The fact that experimental inconsistencies arise so frequently may reflect the complexity of the psychological and perceptual processes involved in the apparently simple tasks of locating objects in the visual field and defining their size and shape. Such complexity also reinforces the idea that the human visual system is not simply a passive recording system but is actively selecting and interpreting visual stimuli in light of context, knowledge, and expectations (Rao and Ballard 1999; Bastos et al. 2012; Friston 2018).

In an attempt to reconcile some of the apparently inconsistent findings in this field, Fortenbaugh et al. (2012) investigated the influence of visual boundaries on the perceived structure of the full visual field. In one study conducted inside a Goldmann perimeter—a device commonly used by ophthalmologists to measure visual field extents—the external borders of the participants' visual fields were defined either by facial features (e.g. nose and brow) or by the introduction of a false boundary consisting of a ring-shaped aperture introduced inside the hemisphere of the perimeter. By using this apparatus, the experimenters were able to avoid the limitations of conventional flat screen apparatus typically used in visual perception studies. Participants were asked to estimate the perceived distance of the targets from fixation relative to the perceived length of either the vertical or horizontal meridian. This task required the participants to first construct an internal representation of the global length of the tested axis in order to generate the judgment. The presence of visual boundaries produced a foveal bias and linear scaling of visual space across all four axes in both monocular and binocular viewing conditions, with a stronger effect in the experimental condition compared to intrinsic facial boundaries. On the other hand, in the absence of external borders (e.g. temporal and inferior axis in monocular viewing conditions), findings reported a non-linear peripheral bias.

What we learn overall from the studies conducted to date on the geometrical structure of the full visual field is that it is highly complex and can vary in a number of dimensions depending on multiple interacting variables. These include the scope and focus of attention (VanRullen et al. 2007), relative luminance of stimuli (Bedell and Johnson 1984), the amount of visual space being measured and its apparent boundaries (Fortenbaugh et al. 2012, 2015). The task of scientifically mapping this complex structure in all its dimensions may be aided by the studies of visual space already undertaken by artists.

7.6   Integrating Artistic and Scientific Mappings of Visual Space

In order to overcome some of the experimental limitations noted above, and to arrive at a more complete analysis of the geometrical structure of the full binocular visual field, we have conducted a number of studies that use a novel combination of methods and techniques drawn from art, computer graphics technology, and psychophysical science. These studies have allowed us to build up a richer and higher-dimensional model of the geometry of visual space than has been available to date.

Our attempts to map the structure of the visual field began by using painting and drawing to map the phenomenology of visual space, including its geometrical structure. Figure 7.7b shows a painting made by one of the authors of a physical space—a

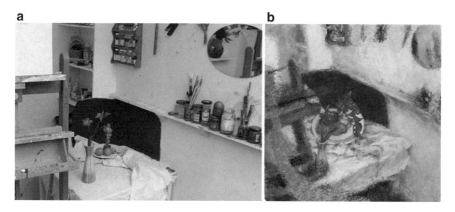

Fig. 7.7 **a** Photograph of a still life scene taken with a standard 50 mm camera lens. **b** *Still life with flowers*, Robert Pepperell, oil and sand on shaped canvas, 2012, 25 × 25 cm. The painting reproduced in (**b**) shows same visual space as the photograph but is organised in a very different way. Note the relative enlargement of the centrally located flowers and the diminution of peripheral objects in the painting compared to the photograph. *Image copyright* Pepperell (2012)

still life scene—and the perceptual structure that was recorded in the painting can be compared to a linear perspective photograph (a) taken from the same view point. To make the painting the author fixated a point in visual space and then visually measured the perceived sizes, shapes and positions of all the objects in the space relative to the fixation point. Judgements about these areas were made on the basis of the information available to the light receptors in the peripheral retina, and in this way the structure of the visual space perceived was mapped. It is notable that the result of this mapping exercise is an image which conforms in many respects to the general structure noted above in the work of previous artists, namely that the central area under fixation (the red flower) is depicted as larger compared to the photograph, and the peripheral regions are more compressed. This results in a painting that is smaller overall than the photograph while containing the same physical space.

To determine whether this same geometrical pattern occurred in the wider population, we carried out experiments in which we asked people to observe an object in central vision and record, using drawing, how peripheral objects appeared relative to the central object (Baldwin et al. 2016). We used the apparatus shown in Fig. 7.8a. The experimental method consisted in asking participants to fixate on the central disc, while paying attention to how each of four peripherally located discs (top, bottom, left and right) appeared compared to the central one. Critically, this procedure was carried out without moving the eyes from the fixation point when recording the appearance of peripheral locations of the visual field. Analysis of the size and shape of the discs drawn by participants revealed that they perceived the peripherally viewed discs as significantly smaller compared to the central one and compressed in shape. The discs physically aligned on the horizontal axis (left and right) were represented as vertically oriented ellipses; while discs aligned on the vertical axis (top and bottom) were represented as horizontally oriented ellipses. Results showed that

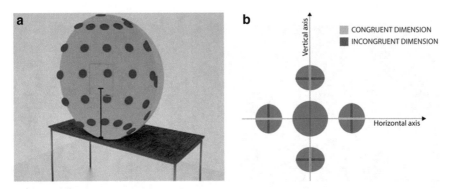

Fig. 7.8 a Shows the hemispherical apparatus used to measure the perceived size and shape of objects in the peripheral visual field. **b** Shows a graphical representation of the perceived size of peripherally viewed discs. The four peripherally located discs are show as drawn by participants, based on the mean height and width values calculated from the drawings made during the experiment. Congruent dimensions are represented in green; incongruent dimensions are represented in red for both orientation axes

participants underestimated the dimension congruent to the axis the disc was aligned with, as illustrated in Fig. 7.8b. These findings showed that participants perceived the structure of their visual field in a way that is broadly consistent with that reported by artists, as described above.

To understand the apparent diminution of objects in the visual periphery at a more fine-grained level, and to test whether similar effects were occurring with rapidly presented stimuli, we created a device that projected computer-generated stimuli onto a curved screen. This apparatus presented stimuli equidistantly from the midpoint between participants' eyes and allowed us to control the eccentricity of the stimuli and their duration (Baldwin et al. 2016). We used the method of constant stimuli to randomly present a set of nine-sized discs, with the smallest being half and the biggest being the double of the size of a central reference disc, at respectively 15°, 30°, 45° and 60° of eccentricity. During each trial, participants were asked to make a simple forced-choice to determine if the peripherally presented disc appeared bigger or smaller compared to the central reference one—which remained constant in size. We performed a Probit Analysis to calculate the psychometrical function for each participant and determine the point of subjective equality (PSE) for each of the four eccentricities. Our results revealed a non-linear trend, with the furthest peripheral location (60°) being the one where participants showed the smallest bias effect compared to the other locations tested. We found that at the closest locations, peripheral stimuli needed to be significantly bigger than the central one to be perceived as having the same size, while at 60° of eccentricity, the size of the peripheral stimuli needed to be much closer to the actual size of the central reference disc to be perceived as having the same size, with a PSE significantly smaller compared to the other eccentricities. Overall our findings were in line with those of Fortenbaugh et al. (2012) and provided confirmation that people tend to underestimate the size of peripherally located objects, for both brief and long exposure times.

One limitation of these studies was that while we were measuring the perceived size of objects across the visual field, including in the far periphery, we were not accounting for the effects of relative depth in visual space. We were aware from the original investigations through painting and drawing that depth judgements are modified by contextual information and peripherally-located cues. In order to measure the geometrical structure of the full visual field in both breadth and depth we constructed a visual scene composed of intersecting metal rods, 100 cm length, arranged as a three-dimensional grid, as illustrated in Fig. 7.9a (Burleigh et al. 2018b; Ruta et al. 2016). The rods were covered in coloured insulation foam, which allowed for easier visual discrimination of the different layers when viewing inside the grid. Within the grid a polystyrene sphere was suspended within the centre of the cube furthest away from the frontal view of the observer. Using this sphere as the fixation point, the observer then made a drawing of the entire visual space from a seated position, shown in Fig. 7.9b. The drawing was not created using any formal mathematical, geometrical or construction method, but instead is the result of empirical observations and judgements arrived at by trial and error using the general artistic method outlined above to measure the full visual field in depth and breadth.

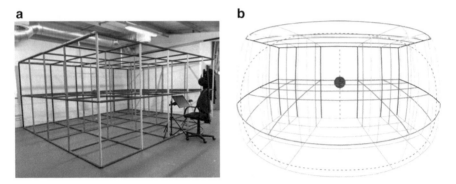

Fig. 7.9 **a** Shows the experimental setup, including the grid room layout, the chair from which the scene was viewed, the location of the computer monitor, and the red fixation ball. Each cube was 100 cm^3, and the room was composed by 5 cubes wide by 4 cubes deep by 2 cubes high. **b** Shows a drawing by one of the authors of the subjective visual experience of the experimental space based on a fixation point at the red ball. The drawing is a synthesis of both eye views, and the dotted elliptical line shows the approximate limits of the visual field

We generated three different projections of the experimental scene: a "natural" perspective projection based on the drawing made by one of the authors, and two standard geometrical projections, a curvilinear fisheye projective and a wide-angle linear projection (see Fig. 7.10). Our study had two experimental conditions: online and in situ (Burleigh et al. 2018b). In the first online study, participants simply ranked the images according to their preference on their computer monitor. In the second in situ study, participants sat in front of the grid room apparatus and compared each of the images we had created with their visual experience of the real scene. For each image, they were asked to fill a detailed questionnaire evaluating how well the picture represented their experience of being in situ, looking at the target sphere. The results of the online study showed that participants preferred natural perspective projection (Fig. 7.10a) significantly more than the other two projections. The results of the in situ experiment showed that the natural perspective projection was also significantly preferred, judged as having significantly higher spatial presence and as being

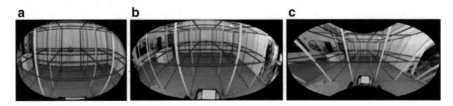

Fig. 7.10 The three projections used to measure the full visual field in depth and breadth, and to quantify which projection was preferred by participants. **a** Natural perspective projection; **b** fisheye perspective projection; **c** linear perspective projection

significantly more comfortable to look at compared to the standard linear and curvilinear fisheye projections. The same result was found for 'perceptual match' ratings, meaning that participants rated the natural perspective image as '*representing exactly how they perceived the environment*' significantly more than the other two projective geometries. Overall, this study added weight to the suggestion that non-Euclidean artistic representations of visual space are judged as aesthetically preferential to standard projective geometries, and match more closely the perceived geometry of visual space.

The final study we report here concerns the hypothesis that the geometry of visual space correlates to the geometrical structure of imaginary space. We were led to consider this hypothesis through analysis of pre-linear perspectival artworks, such as Medieval religious paintings, which displayed some of the same geometric characteristics as the paintings described above, but which were made from imagination rather than observation (Ruta 2019; Ruta et al. 2017). Once again, we used a drawing task as method for recording the subjective experience of participants. Participants were instructed to close their eyes and imagine looking at three identical oranges arranged on a horizontal line. They were then asked to imagine looking at one orange more than the others, first the one on the left, then the one in the middle, then the one on the right. Finally, participants drew what they imagined on a blank piece of paper. Our findings showed that, regardless the position of the orange, participants drew the one they imagined looking at significantly larger than the others (Ruta et al. 2017; Ruta 2019). Some examples of these drawings can be seen in Fig. 7.11. This simple task suggests that the geometry of imagined visual space, as represented in participants' drawings, corresponds to the geometry of visual space, as represented in artistic depictions of real space, where objects under fixation tend to be rendered larger than equivalently sized objects in the periphery.

In summary, the studies reported here provide evidence for the existence of a broadly-defined geometric structure for visual space and the way it is depicted in images. It is a structure that is similar to that frequently recorded by artists, as noted above, in which objects in the central visual field tend to appear larger than those in the periphery. So, although our understanding of the complex and variable structure of visual space remains incomplete, we do have converging evidence from both art and science to suggest that it has certain consistent features. Questions remain, however, about the extent to which a recording of visual space, in a drawing or painting, corresponds to the way space is actually perceived, and why visual space seems to map neatly onto Euclidean space in our everyday visual experience whilst deviating so markedly from it when analysed artistically or scientifically.

7.7   Conclusion

Visual space is a complex, multi-dimensional phenomenon that is yet to be precisely or definitively modelled, either psychophysically or mathematically. The binocular visual field in its fullest extent and in depth are particularly challenging to

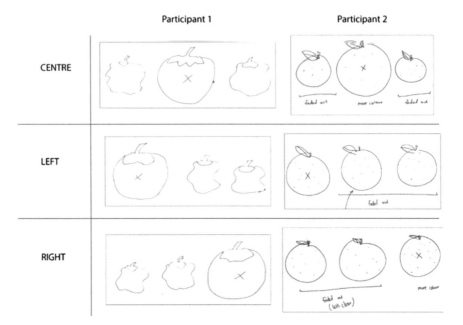

Fig. 7.11 Two sets of drawings made by two participants (one in the left column and one in the right column). The vertical text on the left shows which orange participants were instructed to image they were looking at. Participants indicated the target orange they imagined looking at by drawing a cross on it

measure scientifically with conventional apparatus. Artistic representations of the visual world, however, provide a rich source of objective data about the subjective phenomenon of visual experience, and reveal much about its gross geometrical structure. Our work shows that by combining methods and techniques from art, science and technology it is possible to make progress in studying these complex and elusive phenomena in ways that would not be possible within either discipline alone. The challenge that now faces us is to refine our models of visual space while accommodating all the many factors that influence its structure in order to arrive at a more complete mathematical description of the world we see—as we see it. Such a description would not only be of scientific value but may have useful applications in many areas of imaging technology where it might be used, for example, to improve the way medical images or visualizations of architectural spaces are rendered.

Acknowledgements With thanks to Stephen Gill and Stuart Jeffreys for providing information relating to Rosslyn Chapel and to Jon Clarkson for drawing our attention to Constable's drawing method.

References

Ames, A., Proctor, C., & Ames, B. (1923). Vision and the technique of art. *Proceedings of the American Academy of Arts and Sciences, 58.*

Anstis, S. (1998). Picturing peripheral acuity. *Perception, 27,* 817–825.

Baldwin, J., Burleigh, A., Pepperell, R., & Ruta, N. (2016, July–August). The perceived size and shape of objects in peripheral vision, *i-Perception.*

Bantjes, R. (2014). "Vertical Perspective Does Not Exist": The scandal of converging verticals and their final crisis of perspectiva artificialis. *The Journal of the History of Ideas, 75*(2).

Bastos, A. M., Usrey, W. M., Adams, R. A., Mangun, G. R., Fries, P., & Friston, K. J. (2012). Canonical microcircuits for predictive coding. *Neuron, 76*(4), 695–711.

Behrens, R. (1987). The life and unusual ideas of Adelbert Ames, Jr. *Leonardo (MIT Press), 20*(3), 273–279.

Bock, O. (1993). Localization of objects in the peripheral visual field. *Behavioural Brain Research, 56,* 77–84.

Bedell, H. E., & Johnson, C. A. (1984). The perceived size of targets in the peripheral and central visual fields. *Ophthalmic and Physiological Optics, 4*(2), 123–131.

Bosse, A. (1649). Sentiments sur la distinction des diverses manières de peinture, dessin et gravure et des originaux d'avec leurs copies. Paris.

Bruno, A., & Morrone, M. C. (2007). Influence of saccadic adaptation on spatial localization: Comparison of verbal and pointing reports. *Journal of Vision, 7,* 16.

Burleigh, A., Pepperell, R., & Ruta, N. (2018a). Fovography: A natural imaging media. In *Proceedings of the International Conference on 3D Immersion.*

Burleigh, A., Pepperell, R., Ruta, N. (2018b). Natural Perspective: Mapping Visual Space with Art and Science. *Vision, 2*(2), 21.

Clair, J. (1984). Adventures of the Optic Nerve. In S. M. Newman (Ed.), *Bonnard: The Late Paintings* (pp. 29–50). London: Phillips Collection.

Downes, R. (2005). *Turning the Head in Empirical Space Rackstraw Downes.* Princeton University Press (NY).

Du Fresnoy, C.-A. (1667). *The art of painting.* London: A. Ward.

Elkins, J. (1994). *The poetics of perspective.* Ithaca: Cornell University Press.

Enright, J. T. (1995). The non-visual impact of eye orientation on eye–hand coordination. *Vision Research, 35,* 1611–1618.

Floçon, A., & Barre, A. (1988). *Curvilinear perspective: From visual space to the constructed image.* Berkeley, CA: University of California Press.

Foley, J. M. (1972). The size-distance relation and intrinsic geometry of visual space: Implications for processing. *Vision Research, 12*(2), 323–332.

Fortenbaugh, F. C., & Robertson, L. C. (2011). When here becomes there: Attentional distribution modulates foveal bias in peripheral localization. *Attention, Perception, & Psychophysics, 73,* 809–828.

Fortenbaugh, F. C., Sanghvi, S., Silver, M. A., & Robertson, L. C. (2012). Exploring the edges of visual space: The influence of visual boundaries on peripheral localization. *Journal of Vision, 12*(2), 19.

Fortenbaugh, F. C., VanVleet, T. M., Silver, M. A., & Robertson, L. C. (2015). Spatial distortions in localization and midline estimation in hemianopia and normal vision. *Vision Research, 111,* 1–12.

French, R. (2015). Apparent distortions in photography and the geometry of visual space. *Topoi.* https://doi.org/10.1007/s11245-015-9316-5.

Friston, K. (2018). Does predictive coding have a future? *Nature Neuroscience, 21*(8), 1019.

Gombrich, E. (1982). *The image and the eye.* London: Phaidon Press.

Hansen, R. (1973). This curving world: Hyperbolic linear perspective. *Journal of Aesthetics and Art Criticism, 32*(2), 147–161.

Hauck, G. (1879). *Die Subjektive Perspektive und die Horizontalen Curvaturen des Dorischen Styls.* Wittwer, Stuttgart, Germany: Eine Perspektivisch-Ästhetische Studie.

Heelan, P. (1984). *Space-perception and the philosophy of science.* Berkeley, CA: University of California Press.

Herdman, W. G. (1853). *A treatise on the curvilinear perspective of nature; and its applicability to art.* London: John Weale & Co.

Hockney, D. (2015). *Painting and photography.* London: Annely Juda Fine Art.

Hubbard, T. L., & Ruppel, S. E. (2000). Spatial memory averaging, the landmark attraction effect, and representational gravity. *Psychological Research, 64,* 41–55.

Indow, T. (2004). *The global structure of visual space.* New York: World Scientific.

Kemp, M. (1987). 'Perspective Rectified': Some alternative systems in the 19th century. AA Files, Architectural Association.

Kemp, M. (1990). *The science of art: Optical themes in western art from Brunelleschi to Seurat.* New Haven, CT: Yale University Press.

Kingslake, R. (1992). *Optics in photography.* Bellingham: SPIE Press.

Koenderink, J., & van Doorn, A. (2008). The structure of visual spaces. *Journal of Mathematic Imaging and Vision, 31,* 171.

Loran, E. (1963). *Cézanne's composition.* Berkeley: University of California Press.

Luneburg, R. K. (1947). *Mathematical analysis of binocular vision.* Princeton, NJ: Princeton University Press.

Maillet, A. (2009). *The Claude Glass: Use and Meaning of the Black Mirror in Western Art.* Zone Books (NY).

Mann, S. (1994). Bonnard: Painter of the world seen. *Bonnard at Le Bosquet* (pp. 29–41). London, UK: South Bank Centre.

Mann, S., & Mann, P. (2008). *Sargy Mann: Probably the best blind painter in Peckham.* London, UK: SP Books.

Mateeff, S., & Gourevich, A. (1983). Peripheral vision and perceived visual direction. *Biological Cybernetics, 49,* 111–118.

Newsome, L. R. (1972). Visual angle and apparent size of objects in peripheral vision. *Perception and Psychophysics, 12*(3), 300–304.

Panofsky, E. (1924). *Perspective as symbolic form.* New York: Zone Books.

Paraskos, M. (2010). *Clive head.* London: Lund Humphries.

Parris, L., & Fleming-Williams, I. (1991). *Constable, exhibition catalogue.* London: Tate Gallery.

Parsey, A. (1840). *The science of vision, or natural perspective!.* London: Longman.

Pepperell, R. (2012). The perception of art and the science of perception. In B. E. Rogowitz, T. N. Pappas, & H. de Ridder (Eds.), *Human vision and electronic imaging XVII.* Proceedings of SPIE-IS&T, Electronic Imaging (Vol. 8291, p. 829113). Bellingham, WA: SPIE Press.

Pepperell, R., & Haertel, M. (2014). Do artists use linear perspective to depict visual space? *Perception, 43,* 395–416.

Pepperell, R., & Hughes, L. (2015). As seen: Modern British painting and visual experience. *Tate Papers.*

Pepperell, R., Ruta, N., & Burleigh, A. (2016). Exploring and evaluating a new artistic natural perspective and its possible applications to mindfulness practice [Conference Presentation]. In *Symposium of New Technologies for Mindful Awareness and Wellbeing,* London, November 24, 2016.

Pepperell, R. (2016). Always learning to see: The art and thought of Sargy Mann. *Art & Perception, 4*(4).

Pirenne, M. H. (1970). *Optics, painting and photography.* Cambridge: Cambridge University Press.

Rauschenbach, B. (1982). Perceptual perspective and Cézanne's landscape. *Leonardo, 15*(1), 28–33.

Rao, R. P., & Ballard, D. H. (1999). Predictive coding in the visual cortex: A functional interpretation of some extra-classical receptive-field effects. *Nature Neuroscience, 2*(1), 79.

Richardson, J. (1953). *Georges Braque: An American tribute.* New York: Public Education Association.

Ruskin, J. (1873). *Modern painters* (Vol. 1, revised edition). London.

Ruta, N., Burleigh, A., Vigars, R., Barratt, E., & Pepperell, R. (2016). Evaluating an artistic method for depicting human visual space. In *Applied Vision Association meeting*, London, December 19, 2016.

Ruta, N., Burleigh, A., & Pepperell, R. (2017). Image and imagination: How figure scale in medieval painting reflects visual perception. In Abstracts from the 5th Visual Science of Art Conference (VSAC) Berlin, Germany, August 25–27, 2017. *Art & Perception, 5,* 337–426.

Ruta, N. (2019). *Visual perception in far peripheral visual space and its artistic representations* (Doctoral dissertation). Cardiff Metropolitan University. Retrieved from https://repository.cardiffmet.ac.uk/handle/10369/10944.

Schneider, B., Ehrlich, D. J., Stein, R., Flaum, M., & Mangel, S. (1978). Changes in the apparent lengths of lines as a function of degree of retinal eccentricity. *Perception, 7,* 215–223.

Suppes, P. (1995). Some foundational problems in the theory of visual space. *Geometric representations of perceptual phenomena: Papers in Honor of Tarow Indow on his 70th birthday.* NJ: Lawrence Erlbaum.

Todorović, D. (2009). The effect of the observer vantage point on perceived distortions in linear perspective images. *Perception and Psychophysics, 71*(1), 183–193. https://doi.org/10.3758/APP.71.1.183.

Tyler, C. W. (2015). The vault of perception: Are straight lines seen as curved? *Art & Perception, 3*(1), 117–137.

Uddin, M. K. (2006). Visual spatial localization and the two-process model. *Kyushu University Psychological Research, 7,* 65–75.

VanRullen, R., Carlson, T., & Cavanagh, P. (2007). The blinking spotlight of attention. *Proceedings of the National Academy of Sciences, 104*(49), 19204–19209.

von Helmholtz, H. (1866). *Handbuch der Physiologischen Optik.* Hamburg: Voss.

Wagner, M. (2006). *The geometries of visual space.* Hove: Psychology Press.

Walters, E. (1940). Vision and the Artist, *The Artist,* March 1940.

White, J. (1972). *The birth and rebirth of pictorial space.* London: Faber & Faber.

Yang, Z., & Purves, D. (2003). A statistical explanation of visual space. *Nature Neuroscience, 6*(6), 632–640.

Yu, H. H., & Rosa, M. G. (2010). A simple method for creating wide-field visual stimulus for electrophysiology: Mapping and analyzing receptive fields using a hemispheric display. *Journal of Vision, 10*(14), 15.

Chapter 8
From Sketches to Morphing: New Geometric Views on the Epistemological Role of Drawing

Renaud Chabrier

Abstract Geometry, in the modern sense of the term, is based on the study of transformations. The diversity of those transformations has recently been used to understand the diversity of spatio-temporal representations in the brain. I argue that this new association between geometry and neuroscience can also change the way we consider drawing. The role of transformations is thus evaluated here in three differing examples, all involving the drawing of living bodies: drawing from the nude, drawing a sequence of movements, and life sketching. This analysis indicates that "strokes", rather than "lines", play a fundamental role when drawing living beings. On this basis then, it is possible to put aside the classical approach to dealing with spatial representations through "central" or "linear" perspective, and to highlight alternative principles of spatial composition such as drawing "without a point of view". This attempt to connect drawing with modern geometry and neuroscience leads to a re-evaluation of the epistemological importance of digital animation tools, according to their geometrical premises. I will discuss in more detail the role of morphing-based animation, which is based primarily on texture transformations rather than virtual cameras. The development of such a complete approach to drawing, including all its static, moving and transformational aspects, will emerge as a vital step in addressing the new challenges raised by the representation of life in the twenty-first century.

8.1 Introduction

Geometry maintained an important relationship with drawing for more than two millennia. Then, over the course of the last two centuries, the connection was gradually severed: geometry with all its developments in physics became seemingly too abstract to be studied with the aid of drawing. Indeed, the strange and fascinating

R. Chabrier (✉)
Institut Curie, PSL Research University, CNRS UMR3348, Orsay, France
e-mail: renaud.chabrier@m4x.org

LIX, Ecole Polytechnique, CNRS, IP Paris, Palaiseau, France

© Springer Nature Switzerland AG 2021
T. Flash and A. Berthoz (eds.), *Space-Time Geometries for Motion and Perception in the Brain and the Arts*, Lecture Notes in Morphogenesis,
https://doi.org/10.1007/978-3-030-57227-3_8

aspects of space-time at the scale of atoms or galaxies had little to do with the experience of life at human scale, where drawing practice is deeply rooted.

While these new scientific fields were being explored, major discoveries and innovations were altering the practice of drawing in the Western world: modern art questioned the use of central (i.e. linear) perspective, animated cartoons explicitly introduced movement into images, and new methods introduced by both biomechanics and modern dance modified the way artists were looking at bodies. The discovery of techniques distant in time such as rock paintings, as well as distant in space such as Chinese painting or Japanese prints, also highlighted a certain universality of drawing.

In these circumstances, it may seem logical that a new, more global understanding of the practice of drawing should emerge, in both an international and holistic sense, but this has not been the case. As the philosopher Grossos (2017) indicates, the interest in symbols has led, over a long period, to neglect of the qualities of movement and presence that are fundamental in rock art. Similarly, the interest in central perspective as a symbol of power has greatly overshadowed the analysis of living bodies represented in the same scenes (Chabrier 2016). Indeed, the majority of reference studies such as those of Leroi-Gourhan (1965) on cave art archaeology or Panofsky (1927) on the aesthetic philosophy of renaissance painting prefer to distinguish separate symbolic systems, rather than appreciating the quality of spatio-temporal representations of living bodies in various context.

However, in recent decades, both cognitive science and neuroscience began to explore how the brain deals with space and time. An important idea was introduced: we animals and humans use not only one single universal geometry, but a set of different geometries depending on the task (Bennequin et al. 2009). In this chapter, we will see that this idea opens up the possibility of reconnecting modern geometry with drawing, especially in the case of the representation of living bodies. In the first part of this text, I will present very simply the fundamental role of transformations in the relatively recent re-definition of geometry, then I will discuss the application of this approach in neuroscience. I will propose using drawing both as a means of accessing the different geometries used by the brain and as an effective training tool to make good use of them.

In the second part, we will go on to examine the practice of drawing living bodies by looking for what transformations they imply. We will examine three situations respectively: drawing from the nude, sequence drawing, and life sketching. Each time, I will introduce historical elements to show how these drawing practices have evolved since the late nineteenth century. This analysis in terms of transformations will highlight the fundamental role played by the spatiality of the stroke. I will propose viewing the stroke as an exemplary case of "simplexity", a tool for managing in a simple way a complex problem such as the representation of living beings. Using the word "stroke", instead of "line", will be a key aspect of this text. Indeed, the word "line" has a precise mathematical meaning: a line has a null curvature, otherwise it is described as a "curve". Furthermore, those concepts of line and curve create abstraction from materiality: they do not take into account what the lines and curves are made of. Unlike "line", the word "stroke" is not used in mathematics: it relates to a combination of gestures and traces on paper, with limitless possibilities of both trajectories and material textures. Consequently, speaking of "strokes" instead of "lines" or "curves", when a precise mathematical meaning is not what matters, prevents confusion between different levels of abstraction.

The third part will extend this reflection by examining the way in which several bodies, represented by means of the spatiality of the stroke, can be composed on the same sheet. I will highlight the existence of a geometry "without a point of view" whose use, highly developed in both the arts and the sciences, has been largely overshadowed by the attention paid to central perspective. I will present the technique of morphing-based animation, which can transform, in multiple ways, drawings created using a geometry without a point of view. Morphing will consequently be shown to be a natural extension of drawing, just as 3D computer graphics are a natural extension of both central perspective and photography. From sketching to morphing, this analysis in terms of transformations will emerge as a new overall framework for understanding the epistemological role of drawing. With the help of a philosophical reflection on the respective resources of Eastern and Western art, I will suggest the possibility of new collaborations between production, education and research.

8.2 Multiple Geometries in the Drawer's Brain

8.2.1 Geometry as a Study of Transformations

In the culture of non-mathematicians, the word "geometry" is generally used when speaking of distances and angles between various objects, or between different parts of the same object. The basic elements of geometry taught in school usually involve very static objects in two dimensions such as circles, lines, triangles or more complex polyhedrons. Consequently, the collective and popular understanding of the word "geometry" is still very close to the one expressed by Plato a long time ago: that this science is dedicated to the study of ideal, perfect and unchangeable forms.

In the mathematics community, however, the word "geometry" has taken on a very different meaning over the past 150 years. In 1872, Klein suggested replacing this study of static forms, inherited from Euclid, with another far more general discipline: the study of transformations allowing the passing from one "mathematical object" to the other. Some of those transformations are well known (displacements, also called translations, rotations…) while others are less familiar (affine transformations, perspective transformations…). The important concept is that each type of transformation allows us to associate or to differentiate among the objects that are studied: for example, two triangles presented one next to the other will be said to be equivalent for translation if a simple displacement makes the left one exactly cover the right one (or the opposite). With more complex transformations (rotation, change of scale, shearing…), mathematicians are able to create various classes of equivalence between the objects they are interested in, some of them being more intuitive than others. According to Klein's approach, then, a "geometry" is defined by what one can do with a certain kind of transformation; thus, there is not one, but many geometries.

Clearly, a deep epistemological gap lies between those two notions of geometry: one is "unique" and is based on unchangeable forms, the other is "multiple" and seeks possibilities for transformation. In the twentieth century, replacing the existing geometry with a more "modern" one in basic education was a fundamental challenge. However, despite the efforts of teachers to bring about this change (especially in France during the second half of the century), one must admit that the modern meaning of the word "geometry" was never adopted by the general public. Meanwhile, during the same period, the basic teaching of classical geometry progressively weakened, even in scientific and technical programs. One can remark that this double failure coincides with the neglect or even the rejection of the use of drawing in the teaching of mathematics, well before the more recent and radical change in representation practice due to digital tools. This separation between drawing and mathematics is partly due to the fact that drawing has long been seen as a "static" and "realist" representation tool in Western society. Thus, it seemed fundamentally incompatible with the development of a new concept of abstract transformation. Nonetheless, we can observe that a number of mathematicians make effective use of hand-made drawings when explaining their research to a wider audience, such as Patrick Popescu-Pampu when he addresses the relationship between topography and geometry (2018; see also online resource [b]).

8.2.2   The Brain Can Use Multiple Types of Transformations

Whatever the discussion of the teaching of "geometrie-s" in education may be, we can observe that a perfectly normal brain without special mathematical training uses at least three geometries. They are based respectively on simple displacement (translation), scale change, and rotation. These three "basic" transformations allow us to judge that several forms are not only alike, but alike in different ways and at different levels. Today, such transformations are widely used in modern animation and motion design, since animation software makes them very accessible. Before the widespread use of digital tools, however, artists like Norman McLaren showed the expressive richness of animations rigorously limited to a few transformations of this kind in short movies like "Vertical Lines" (1960). Similarly, psychologists have used very basic animations, such as moving triangles, in order to study the tendency for humans to assign a personality to moving objects (Heider and Simmel 1944; White et al. 2011). Clearly, the conceptual facility of displacement, rotation and (small) scale change has allowed a natural connection between animation and cognitive psychology, both before and after the introduction of digital technologies: animated stimuli played an important part in the study of how the brain deals with transformations.

Owing to a long philosophical tradition going back to Aristotle, movement is generally considered to be a secondary attribute of form: you create a form, and then you can move it. However, the practice of animation enables a conceptual reversal, very similar to the one that we just discussed in mathematics. Indeed, in an animated sequence, it often appears that forms exist for the viewer mainly through their movement. Another way to describe this is to say that the forms presented to the audience remain invisible (or rather, neglected) if they are not carried by a work of transformation, which reveals them and orients the way they are perceived.

Transformation are present everywhere in animation. "Simple" parameters of position, size and rotation can be considered as fundamental transformations, especially in the context of documentary films. They are intended to be used with subtlety, in order to induce various evolutions in the imagination of the audience: indeed, a simple zoom on an image, centered on the right place with the right dynamics, can alter the intelligibility of a sequence. The short film "Birth of a Brain" (Fig. 8.1 and Movie 2016) is a good example. It aims to make brain development accessible to a wide audience by showing the richness of the associated spatio-temporal phenomena. A wide range of transformations are used in this film, including morphing, which we will discuss later. Many of them are based on the basic operations that we have just described.

While those very simple (but very efficient) examples do not challenge the geometric world we are familiar with, it appears that other more unexpected geometries also co-exist in the brain. This idea has been developed from studies of "simple" curved movements in tracing tasks, and studies of trajectories in locomotion tasks. Bennequin et al. (2009) found that such movements could be transformed one into

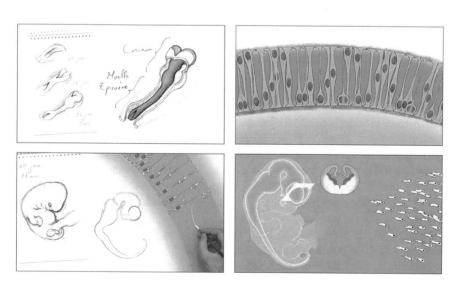

Fig. 8.1 Showing transformations of the brain with transformations of drawings. Images from "Birth of a brain", directed and drawn by R. Chabrier for the "Brain" exhibition at Cité des Sciences et de l'Industrie, 2014

the other according to a number of specific geometric rules. Bennequin and Berthoz (2017) also proposed extending this approach to more complex movements like prehension, with more advanced mathematical tools. Henceforth, the "objects" that those researchers are interested in are no longer static forms, possibly animated, like squares or triangles. Those objects are in fact defined in time and space: they can be in motion, and the geometries concerned can even be used to represent forces.

The rigorous expression of these last mathematical ideas is very abstract, and somewhat inaccessible even with a scientific background, such as the "geometries without points" of Elie Cartan (Ivey and Landsberg 2003). Moreover, those concepts are applied to neuroscientific observations, which can also be difficult to understand. In addition, animation techniques that use more complex transformations than displacement, rotation and scale change require a distinct professional training, incompatible with a scientific career. Thus, the connection between animation and science has become quite strained. But if the brain indeed works this way at the functional level, we can infer several consequences.

8.2.3   Drawing as a Way to Access to the Diversity of Geometries

First, it should be possible to detect the influence of those multiple geometries in many human activities, and perhaps find examples that are easier to understand, to experiment with, or to sense, than their mathematical expression. Drawing makes a very good candidate for this investigation, because it can change both "three-dimensional reality" and "abstract thinking", into fascinating fields for geometrical explorations. Escher's etchings already provide well-known examples of this fact in imaginary worlds, but further on we will see that interesting geometrical challenges can be found in the reality in front of us, as soon as we pay attention to the spatiality of living beings.

Second, it may be necessary to assume the geometrical richness of human activities, in order to better evaluate the benefit that the brain derives from their practice. In particular, many manual activities related to "arts and crafts" like wood-working or modelling, are better appreciated today for their balancing function. In contrast with "pure" digital practices, they involve gesture, in which trajectories and forces are closely embedded. Adopting Daniel Bennequin's proposition, we can consider those actions to be necessarily related to more complex geometries than the Euclidian one, integrating forces in particular.

Once again, drawing can play here an interesting role of mediation between the concrete and the abstract. Indeed, while aiming at the work of representation prior to production, fabrication or construction, the materiality of the paper and the tools (pen, charcoal, black stone, brush…) implies a play of forces between thrust, traction and friction, which must be partially integrated in planning a gesture.

While the proposition of Berthoz, Flash and Bennequin initiates a new way of considering spatio-temporal representations in the brain, it also encourages a change in the way we consider the activities that generate those representations, which can lead in turn to a fundamental re-evaluation of manual craft. If we follow the logic of the authors, initially proposed by Klein, the key lies in the possibilities of transformation. Consequently, we will now take a look at drawing in this way, focusing, at first, on sketching living beings, then on composition, and finally on morphing-based animation, in order to understand which geometries arise from those practices.

8.3 Drawing Living Beings, from Transformations to Strokes

Fundamentally, drawing is a temporal process. It simply cannot be idealized as an instant event. This clearly distinguishes it from photography, which is generally based on extremely rapid shooting. As soon as one tries to draw living beings that can move and change, the time taken by the act of drawing opens the door to many transformations, as much in the subject drawn as in the draftsman himself. However, these transformations will appear very differently depending on the contexts in which the drawing is practiced. I will now review drawing from the nude, drawing in sequence, and life sketching. As we will see, those three approaches to representing life through drawing are related to different epistemological backgrounds, that may hide or highlight the way transformations are involved. Thus the main purpose of this section is to clarify this situation. It will become clear that the concept of "stroke" provides a more accurate basis than the usual concept of "line", if we are interested in thinking about drawing in terms of transformations.

8.3.1 Suspending Transformations: Drawing from the Nude

The nude illustrates well the temporality of drawing. A session usually lasts three hours, during which the model adopts poses of different durations (2, 5, 10, 20, 30 or 45 min …). During each pose, the models focus on suspending the transformation of their body, in order to let themselves be drawn. But throughout these three hours, many other transformations occur in both the draftsmen and the models: the postures are relaxed or refined, the breathing is stabilized, time takes on a different density, the hand adapts to the attrition and sharpening of the tool, the drawing itself evolves. These are "invisible" transformations that are well known to trained draftsmen and models, but are at the same time difficult to describe and analyse.

Once the session is over, the spatial presence of the model, translated by the drawing, remains on the sheet of paper. This presence can also be interpreted in terms of transformations: the drawing suggests what would happen if one started to turn the drawn body around, or if this drawn body started to move. However, this way of seeing requires some training, and some sensitivity to the anticipatory processes involved in the brain. In the example that I give (Fig. 8.2a), it can be seen that my drawing evolves as the session progresses. In addition, the first relatively short poses may suggest a movement of the model, whereas a longer pose, drawn from two different angles, gives an idea of the model's spatiality through a turning movement of the viewer, if one passes the eye from one drawing to another.

In general, for cultural reasons that are deeply rooted in Western society, drawings produced from the nude are rarely interpreted in terms of transformation. Rather, they are considered as "still images" fixing the presence of the model on the paper. This stillness was indeed deliberately sought in academic drawing in the eighteenth and especially in the nineteenth century, treating living bodies as marble sculptures. As evoked by Bourdieu (2013) in his sociological study of the pioneers of modern art, the organization of life drawing workshops in France in the nineteenth century imposed a very strict hierarchy amongst the students, using geometrical constraints: on the one hand, they were asked to draw by rigorously obeying the laws of central perspective, and on the other, only the highest-ranked students were permitted to choose their "point of view". From the 1860s, Manet and subsequent modern artists began to reject this form of drawing, since they considered it both fixed and constraining.

Today, therefore, for both cognitive and historical reasons, the existence of a strong geometric relationship between a drawing and a "still" model can easily lead to confusion. Firstly, it leads to the idea that the drawing "is" the thing it represents. This confusion related to identity is similar to what happens with photography, especially when it comes to very fine greyscale works with pencil. Secondly, the apparently direct link between the drawing and the model encourages the idea that the drawer, stroke by stroke or line by line, reproduces only "what he sees". This interpretation does not sufficiently emphasize the process of transformation that must be gone through in order to find the coherence of a set of strokes, and it inhibits the imagination of possible transformations from the drawing as it is made. In fact, drawing from the nude can be connected to and practiced with a richer understanding of geometry, but in order to understand how, we might first consider other ways to draw life, further from the academic world.

8.3.2   Developing Transformations: Sequence Drawing

Sequence drawing consists of drawing several steps in a process: most of the time, the steps are close together, in order to make the evolution more evident (especially if the sequence is not displayed as a movie, but printed on paper). Consequently, as opposed to drawing from the nude, sequence drawing is explicitly connected to transformations. For example, a sequence can show the development of the embryo, or a running rabbit (Fig. 8.2b, c). In the first case, modifications of the drawn body are particularly important. This corresponds to the meaning we usually give to the word "transformation". However, we must remember once again that in mathematics

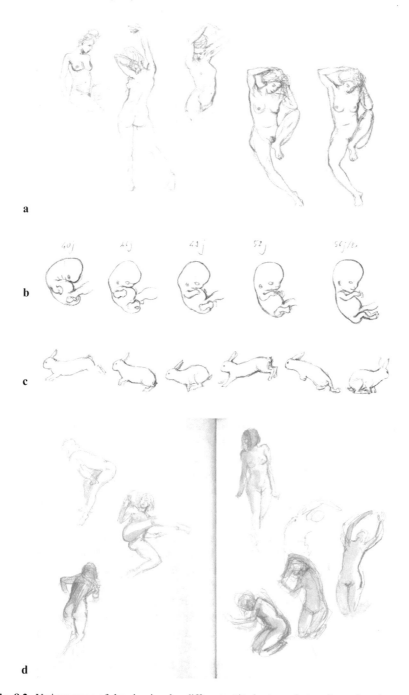

Fig. 8.2 Various ways of drawing involve different attitudes towards transformations in general, and body movements in particular. **a** Drawing from the nude asks the real model to suspend his/her transformations for a few minutes. **b, c** Sequence drawing develops the transformations of an imaginary body in an ordered, analytical way. **d** Life sketches let the model or subject free to transform itself, and use this information for free composition. Drawings by R. Chabrier

this word also refers to modifications that do not affect the whole body shape and proportions; simple translations, changes of orientation and articulations of the limbs, that is to say everything we usually call "movement" or "motion", can also be seen as different types of transformations.

These two drawing sequences are extracted respectively from a scientific film for the general public ("Birth of a brain", Movie 2016), and from projections accompanying a musical comedy ("Snow White" directed by Grimbert et al. 2013). Although they have been produced in order to be shown in totally different contexts, they have much in common. Critically, neither faithfully reproduces a "model" existing in reality; whether it is the embryo or the rabbit given as an example, they are created by drawing inspiration from several sources at once.

From an historical point of view, it is interesting to note that these two types of drawing "in sequence", for either scientific or entertainment purposes, were initially developed during the same period: Haeckel popularized the depiction of embryonic development through his famous drawings in 1874, while Emile Reynaud met success with his first animated cartoons in 1877, thanks to his praxinoscope. However, despite the fact that scientific drawing and animated drawing generated a common and simultaneous interest in transformations, creating a relationship between the two disciplines was not straightforward at the time. Indeed, the idea of drawing the sequential development of a transformation was not received in the same way in the entertainment industry and in life sciences. Imagination was highly valued for entertainment. Conversely, Haeckel was criticized for introducing imaginary aspects into the sequential development of his drawings.

To understand this difference, it is helpful to go back to the historical and epistemological study of the notion of "Objectivity", conducted by Dalton and Galison (2007). In particular, the authors describe how contemporary scientists reproached Haeckel for being overly guided by what he knew, or what he thought he knew, instead of recording correctly what he saw. In general, from the end of the nineteenth century, many scientists actively tried to reject any personal interpretation. Indeed, they were increasingly interested in phenomena that were not easily perceived by the human eye, and they realized that drawing from direct visual observation could transmit many interpretation errors, such as overlooking the diversity found in snowflakes or perceiving a non-existent symmetry in falling droplets. This led those scientists to reject drawing in favour of photography, in order to attain "mechanical objectivity". Still images, in the literal photographic sense, became the main means of studying movement in science.

Hence, there was an early divergence between the art of animation and the sciences of movement, or transformations in general. Animation movie directors wanted their draftsmen to move everything that they could move, and let their imagination run free. Conversely, most scientists wanted to eliminate, as much as possible, human imagination from any visual information created. Animation continued to rely on manual craft to a large extent, while scientists began to value images obtained with automatic and mechanical recording tools. Ultimately, for many scientists in mathematics or physics (but also later in biology), the preference was to have no image at all, since images could interfere with essential concepts.

However, as distant as their practice could be during the course of the twentieth century, animation and science shared a common concern for an aspect of transformations that I have not addressed until this point: invariance. Invariants are aspects or properties of objects that remain unchanged by transformation. In mathematics, Klein's new way of studying geometries was based on such invariants, which were considered as characteristics of the various kinds of transformations. In an animated fiction movie, invariance is also an important issue: some transformations might affect the very identity of a character, while others will allow you to believe in its integrity from one frame to the other. For example, if we consider again Fig. 8.2b, c, those transformations are intuitively related to the same embryo or the same rabbit.

Such an assumed invariance of a character, in a sequence of drawing that shows a movement broken down into a few steps, suggests that this transformation is in fact continuous. In mathematics, this would mean that analytical tools can be used. But in drawing, too, the realization of such sequences requires a strong "analytical" component: in order to ensure that the views are correctly linked to each other, the eye of the draftsman constantly compares, differentiates and regulates the steps. Very often, in order to support his strokes, the draftsman (or the animator) uses simple trajectories or volumes whose transformations are easier to control. This realization process, which can be very conceptual, brings us closer to the idea that the draftsman draws "what he knows" much more than "what he sees".

In reviewing the cases of drawing from the nude and drawing movement sequences, my purpose was, first, to present two different approaches for drawing living bodies, and second, to show the reasons why one can so easily be reduced to "a pure perceptual product", while the other can most often be interpreted as "a pure conceptual product". As Seymour Simmons has shown in his historical study of drawing instructions (2011), both these ways of considering the practice of drawing have a clear philosophical background. They are closely associated, respectively, with the "realist" and "rationalist" schools of thought, which competed with each other during the nineteenth century and still remain influential; in British and American culture, for example, scientific drawing is still associated with the realism of Ruskin, with an emphasis on perception, while in France the importance of rationalism led to the early decline of drawing in many scientific fields, in favour of conceptual schematics.

According to Seymour Simons, both perceptual and conceptual approaches to drawing tend to neglect the haptic dimension of this practice, and more modern approaches to drawing can be related to "pragmatism". Typically, in the drawing instructions provided by Nicolaides (1941), the gestures of the draftsman can be easily identified, and the strokes produce not only the shape of the bodies, but also aspects that are more difficult to see such as energy or action lines running inside the body. In the following section, I will present a slightly different approach to drawing, similar in the sense that it cannot be reduced to "what I see" or "what I know", but different in the sense that it will not focus on gesture expressivity. Rather, in examining life sketches we will be largely concerned with the relation between strokes and transformations.

8.3.3 Getting Inspiration from Transformations: Life Sketches

Life sketching simply consists of drawing living beings that do not pose. It may be as well to sketch animals, as to work with a model in motion (Fig. 8.2d), as Rodin experimented with from the 1880s. In such cases, all transformations are possible (let us repeat that we use "transformations" in the broad sense of the term, including very simple "movements"). This freedom, left to the subject, offers a great wealth of information on body structure, and it often reveals forms more clearly than the static observation of a motionless body. However, this information, by nature, is accessible only during the time of movement, or change. It has to be stabilized and completed with other information, in the brain. This situation forces the draftsman to acknowledge that the drawing action is located largely outside the strict limits of the visual world.

Suppose for example that I find myself in front of a group of flamingos in a zoo (Fig. 8.3a). I start to draw a bird with its head down, but it then raises its neck. In order to go on, I have to use my short-term memory as well as my anatomical knowledge. More precisely, I have to fit this anatomical knowledge with what the flamingo does: this can be considered as a transformation. I also have an interest in taking into account the "embodiment", which allows me to evaluate the animal's balance by appealing to my own postural sensations. Once again, it is a kind of transformation from one body to the other. In parallel, I must keep in mind that the real bird, although it has moved, continues to be a reference with which my drawing

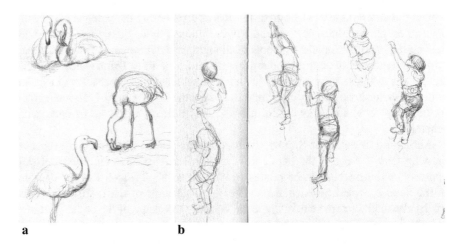

a   b

Fig. 8.3 Life sketches, in the case of animals or children, highlight the relationship between transformations and composition. **a** Different drawings of the same flamingo at different times can also be interpreted as different flamingos within the space. **b** The composition of several individual sketches generates a spatial representation at a larger scale. A unique point of view cannot be identified in such a composition. Drawings by R. Chabrier

can remain coherent in one way or another. Through all those transformations, a vast combination of information and representations (visual, haptic, proprioceptive …) can be mobilized. At each stage of the drawing, all these processes lead to a simple stroke.

Thus the psychological foundation of the stroke represents a rich epistemological issue. In life sketching, the stroke is evidently not the result of a planned action with reference to a single fixed model, whether located "in front of the draftsman" or "in his mind". On the contrary, transformations seem to be involved in the production of the stroke at all levels of abstraction that the brain allows. Consequently, the stroke can integrate and transmit a wide range of information. In this way, it can be considered a case of "simplexity": a tool that makes it possible to handle great complexity in a simple way, without reducing this complexity to the point where oversimplification introduces new problems (Berthoz 2009).

Indeed, a stroke is always simple, compared to the reality, and compared to the neural processes that motivate it. Simultaneously, the effective stroke in a sketch task is never so simple as to be reduced to a two-dimensional "graphic object". In the case of life sketching, it is particularly inappropriate to ignore the spatiality of life: flat sketches lose the connection with reality, and very quickly become boring. Therefore, the stroke that synthesizes this wealth of information must also have a spatial value. This is obtained in many ways: not only by the path of the tool on the surface of the sheet of paper, but also, more precisely, by the variations of curvature, thickness and texture along this path. Those textures, which involve spatial frequencies, also reinforce the possibilities of connections between multiple strokes. Those relationships have a temporal dimension, since it takes a certain amount of time to appreciate each texture. All these factors depend on the force and friction applied in the contact between the tool and the sheet. They bring to the stroke additional dimensions allowing the observer's brain to make a relatively clear qualitative distinction between a flat "graphic object", which can be easily identified and isolated, and a "stroke", which generates space and time, and which binds to other strokes in specific ways.

8.3.4 Intermediate Conclusion: The Geometries of Strokes

I have highlighted the concept of "stroke" (in French, "le trait") through the case of life sketches. However, one can observe that strokes are also involved in drawing from the nude or drawing in sequences. This community of means is evident when drawings are kept as sketches rather than being developed into "finished" images. In Figs. 8.2 and 8.3, I have purposefully chosen examples of this kind. If strokes are deeply related to transformations, as I have attempted to demonstrate, it follows that the geometries of strokes have a fundamental role in drawing.

However, this idea is quite novel in Western culture. A general preference for "finished" and "clean" images, rather than sketches, may explain why the concept of strokes was not greatly developed until recently. The variety of tools (charcoal, pen, pencil, black chalk, blood, wash, not to mention the different engraving techniques) was probably also confusing, since each technique has its specificities when it comes to generating space from a set of strokes, as evidenced by the drawing treatises of the nineteenth century (Fraipont 1897). In any case, the standard concept associated with drawing has not been "stroke" but "line". This vocabulary had huge consequences for the place of drawing in modern society, because it led to the development of a flawed relationship between drawing by hand and mathematics. Indeed, a "line" is by default considered a "straight line". Of course, a line can be curved. But being curved, according to this logic, is only a secondary attribute.

Conversely, using this idea of "stroke", one can develop an alternative concept where curvature is fundamental. This creates a natural connection with the immense developments achieved in mathematics since Klein. Hence, in her work on the analysis of human arm movement, Tamar Flash uses the moving frame, an analysis tool developed by Elie Cartan, in order to show that equi-affine geometry is more relevant than Euclidian geometry for understanding human tracing gestures (Flash & Handzel 2007). This sentence might of course mean nothing, without further explanation, for a non-mathematician reader, but Tamar Flash's further conclusions make a great deal of sense to people who practice drawing; in the affine geometry that she associates with strokes, the most natural connection between a departure and an arrival is not a straight line. This is exactly what happens when we draw living bodies (as opposed to drawing classical or modern architecture): if we have a clear feeling of what we draw, of how the pen starts and of how it ends, our hand simply doesn't take a straight path. Moreover, Tamar Flash defines an affine distance all along this path; she shows that the preferred curves, in human gesture, maximise this distance. Once again, this appears very consistent with the actual experience of drawing: draftsmen tend not to rely on such a feeling as minimizing a distance, using "dry" and straight segments of lines. On the contrary, appropriate strokes provide something more like a subtle nourishing feeling, related to some kind of generosity from which we can benefit, wherever it comes from.

Independently of this very new scientific means of access to the geometries of strokes, this concept already has a rich history outside of Western culture. Chinese painting (which is more precisely translated as "drawing with a brush" according to Fong 2003) relied essentially on brush and ink for many centuries, perhaps making it easier to consider the concept of stroke (Fig. 8.9). Shitao, one the most famous Chinese painters, even wrote a treatise where the concept of stroke plays a major role (Shitao 1710). Of course, accessing this information in translation from the original language brings its risks, but many of those issues have now been addressed, particularly by the philosopher François Jullien. Considering the importance given to transformations in ancient China (Jullien 2009), this culture is worth considering here. Since the Chinese concept of stroke has now been made more accessible, what can we learn from it?

Two important aspects of strokes, addressed by Shitao, are worth highlighting here. The first is the importance given to the "empty wrist", when strokes are produced. This may give the impression that freedom of movement is completely embedded within strokes, but we must take care not to interpret this too quickly as "complete freedom in an Euclidian space": in the next part, we will see that simple composition processes in sketching suggest other kinds of spaces, with other kinds of freedom. A second important aspect of drawing with a brush, according to Shitao, is the distinction between "brush" and "ink" as two components of strokes. Both of them are equally important, and they act in a complementary way: "spirit dimension" is given to the ink by the movement of the brush, and ink provides "animation" to the brush by the use of different levels of darkness (Jullien 2003). While remaining very cautious when it comes to adapting such ancient Chinese concepts in the present, this association between "brush" and "ink" can help us to escape from very deeply rooted habits when we address questions related to strokes, drawing and representation in general. Indeed, we are used to thinking in terms of "form" and "texture", and to considering that the latter is a secondary attribute to the first. Typically, in computer graphics, abstract surfaces will be defined first, followed by textures. In the next section, when I introduce the animation technique of morphing, I will show that a more balanced relationship is possible.

8.4 The Geometries of Drawing

The first part of this chapter outlined how various transformations, as defined in modern mathematics, could be integrated into the way the brain works. The second part questioned how different transformations of the aspect of living bodies could be taken into account in the practice of drawing. This reflection led us to considering the geometries of strokes. This final part will address the possibility of transforming the drawings themselves, and consequently accessing some of the geometries of drawing in a more general fashion.

In the first instance, I will stress the notion of composition, in the process of life sketching. I will infer the possibility of using a geometry "without" a point of view. This geometry will be mainly characterized by animation techniques such as "recomposition", "travelling" or "scale change", in an attempt to adapt the logic of mathematics to the domain of scientific drawing and film-making. Then I will suggest that a more advanced animation technique, known as "morphing" might provide us with deeper insight into the geometries of drawing.

8.4.1 From One Body to the Other: Composition with Space

Sketching living beings is a very ancient activity. For millennia, millions of draftsmen have been sharing a very simple problem: that of how to compose several views of living bodies on the same surface. Life sketching is a good example of this situation. Suppose, for example, that you are sketching in the presence of a living, moving subject. Say, a group of young children during a climbing lesson (Fig. 8.3b). In such a context, it is helpful not to draw too big, since your memory does not necessarily allow you to develop a full page drawing. Furthermore, you don't want to use up too much paper on quick sketches. Consequently, it seems logical to draw one action here, and another action there, on the same page. Let's consider what happens when you draw first one child, and then what happens when you draw a second one (which may be the same child, in another situation).

In order to sketch children, you must certainly draw quickly, but speed alone will never be enough. The moment your pen touches the paper, you can be sure that what you just saw will have changed. So, you must use complementary strategies. Instead of beginning hurriedly, you can refine the way you look, in order to memorize aspects that will help you go on drawing what you wanted, when it is no longer visible. A knowledge of functional anatomy is very valuable in such cases: it allows you to feel how the body organizes itself in space, in relation to the physical surfaces on which forces are produced. Generally speaking, any consciousness of the structure of what you draw, whether it is a child or a mountain, will help you a great deal. Before the first stroke has even been drawn on the paper, it can be backed up by a whole spatio-temporal understanding of the situation, which can guide the following sequence of strokes, resulting in the drawing of one child, doing a certain action, on the paper.

Then, you can go on and make a second drawing. At this precise moment, a very interesting and fundamental aspect of drawing emerges: the second drawing can be made in relation to the first. More precisely, the spatio-temporal understanding that I just described can be used to find an interesting place for the second, not only in relation to the frame of an empty page, but also in relation to the first drawing. Then the third drawing can be made in relation to the former two, and so on.

In general, whether the draftsman is working from a fixed model, from a moving model, or from his imagination, the process we have just described is always possible: it simply adapts to different durations. This composition work seems simple, maybe even trivial. However, it is far more than what today we would call a "graphic composition" in two dimensions, because the drawings, like the strokes they are made of, are not "flat graphic objects". Each of them generates its own spatiality, and this spatiality can be connected to the spatialities of the others. Thus, in order to find a proper place for each new drawing on the page, one has to evaluate the spatiality involved in the former ones, and make choices in order to adapt the new drawing to its context. All this mental effort relies on the capacity of our brains to imagine transformations. Now we will look at how this composition process might be used to characterize an unexpected geometry in the brain, with a direct application in the epistemology of representations.

8.4.2   Geometries with or Without a Point of View

The compositions of life sketches described above highlight a very interesting characteristic, which has so far gone largely unremarked: although the compositions have a highly developed spatiality, they do not use a "point of view". Indeed, in such compositions, it is impossible to say that the "observer" was "here". It is certainly possible to say *from which direction* each body was "represented" or "viewed". But it is simply not possible to use this information to define a particular point or place from which *the entire scene* has been viewed. In a similar way, this type of drawing clearly escapes the notion of "instant", since it does not represent an action taken at a particular moment: it opens up instead a range of possible temporal interpretations. Compared to the geometry "with" a point of view, as will be further evidenced, the geometry without a point of view prevents the image from being identified with what a viewer can see from a particular place, at a precise moment.

Such a geometry has been extensively used historically. Cave art, in the grotte Chauvet or in the grotte de Lascaux, shows this kind of composition. Much later, Chinese painters mastered this absence of point of view. Interestingly, this may perhaps be linked to the fact that the verb "to be" does not exist in the Chinese language. Indeed, in Western philosophy and aesthetics, the verb "to be" plays a central role in the unending quest for "identity" and "essence", with unbroken reference back to Aristotle. The epistemologist Alfred Korzybski characterized this use of "is", in the form of an identification between two concepts, when we say "something *is* something else". He showed that this way of speaking tends to induce confusion between levels of abstraction (Korzybski 1933). In our case, the identification of a painting with the particular object that it represents, or with the point of view that it uses in time and space, is strongly related to a philosophy built around the verb "to be". This might simply have no meaning in the Chinese cultural context, before the end of nineteenth century.

The nineteenth century was an extremely rich period of cultural exchange between East and West, whether welcome or not. While China was discovering some of the new constraints of Western civilizations, including central perspective, Japanese artists like Hokusaï produced renowned collections of woodcuts, where compositions are clearly based on a geometry without a point of view (Fig. 8.4c). Such woodcuts became an important influence on Western modern artists such as Monet or Rodin, who were beginning to break free from the established academic rules.

However, geometry without a point of view was not unknown to Western art at all. In fact, Leonardo da Vinci shows a thorough practice of it when he works on cats, horses or workmen (Fig. 8.4a). The drawings and paintings of Hieronymus Bosch are evidently based on a similar process (Fig. 8.4b). In this type of drawing, the direct observation of successive moments is often combined with imaginary developments on the same subject. To appreciate them properly, we must pay attention to all the possibilities of spatial and temporal combination between bodies. Depending on whether he attributes the different sketches to a single body seen at different times, or to several bodies that coexist, the spectator's gaze can appreciate the passage from one posture to another, or alternatively, he can travel in a common space.

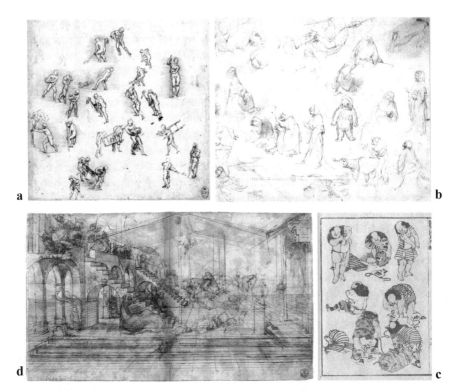

Fig. 8.4 Geometry without a point of view has been extensively used in both occidental and oriental art, typically for the study of living beings. **a** Leonardo da Vinci, movement study with men at work, 1506–1508. **b** Hieronymus Bosch, various figures study, mid-16th century. **c** Katsushika Hokusaï, Manga vol. 9, 1819. **d** Geometry with a point of view, i.e. linear perspective, has been used alongside by Renaissance artists for other composition purposes. Leonardo da Vinci, linear perspective study for The Adoration of the Magi, 1481

As we have seen, a geometry is defined by a set of possible transformations. Geometry "without a point of view[1]" offers considerable possibilities for recomposition by simple displacement of the represented bodies: a double-page of one of my own sketchbooks can be recomposed in another format (Fig. 8.5). This last transformation does not allow any random reorganization, however: one must be careful to maintain a sufficient dialogue between the spaces occupied by the characters. It is also possible to move your gaze onto the composition, to bring it closer or move

---

[1] The reader with a background in mathematics might think, with reason, that what I am describing here is simply linear perspective with a point of view at infinity, or parallel perspective. This kind of perspective is largely used to produce very neutral representations of space in an engineering context. By introducing a "geometry without a point of view" my intention is to stress the composition freedom it provides and to stay close to the experience of drawing, rather than to identify this process with a special case of central perspective. Indeed, fundamental aspects of drawing tend to be discarded when the idea of "point of view at infinity" is introduced, especially texture, which is strongly related to proximity.

Fig. 8.5 Compositions without a point of view can be easily recomposed, while maintaining or enhancing their spatiality. Drawing by R. Chabrier

it further away, to focus or expand, without losing the spatiality it offers. All these observations would of course be trivial if we were interpreting the composition in a two-dimensional space: they become interesting if we adopt a spatial interpretation of the drawing.

A great paradox, as far as Renaissance artists are concerned, is that they are often considered as discoverers and promoters of "the" central perspective, while they were actually mastering both techniques: using a point of view, or not. Indeed, geometry "without" a point of view can be compared to geometry "with" a point of view when comparing two drawings by Leonardo da Vinci (Fig. 8.4a, d). In the first case, as we have already seen, the space is structured by the relation between the horse bodies. In the second case, the space is primarily structured by a grid. In this instance, we can say that bodies do not structure space: they are placed, in the second stage, "in" a well-defined three-dimensional space that pre-existed them.

It can be noted, in the drawing with the grid, that the lines converge in depth towards a vanishing point. The existence of the latter is due to the laws of projection of a scene on a plane, from a particular point of view. Opposite the vanishing point is thus a privileged "place" for the draftsman and the spectator, which is simply the place that provides the best illusion of space (similarly, there was once a specific place for the king in theatres of the eighteenth century, which determined the design of the scenography). According to their place relative to this privileged observer, the representations of the different visible objects will then undergo specific deformations. Thus, in theory, objects drawn in the context of a central perspective cannot simply be relocated elsewhere: we must apply a new perspective transformation to adapt them to their new place. Similarly, if we want to pan above the picture: we should be able to update the entire transformation in order to maintain the illusion of depth in the context of central perspective.

Consequently, operations such as "recomposition" and "travelling" are more complex if a point of view is introduced into the geometry. The increase in complexity is even greater when it comes to "scale change". This has huge consequences for the range of transformations that the brain can interpret, when it processes an image. In a previous study, I demonstrated that the absence of central perspective was fundamental in scientific drawing, since this geometry would limit the required imagination (Chabrier 2016). Since the present study is dedicated to space–time representations, I will focus here on the issue of animating drawing: in this case, the imagination of transformations in the brain has to be converted into methods and algorithms, in order to produce an animation that provokes the same effects (recompositions, travelling, scale change, or more complex phenomenon) in the brains of as many people as possible.

8.4.3 Transforming Drawings: From Compositing to Morphing Animation

Animated movies can use many different media: drawings but also clay, paper cuts, and so on. Here, I will focus on animated movies which are mainly based on drawings (or paintings) that were initially made on a physical background. In this instance, the animation consists of transforming those drawings. Consequently, knowing if those transformations are related to a geometry with or without point of view is essential, since this will alter the effect of the animation in the viewer's brain.

As we have already seen, the absence of point of view appears natural in scientific drawing. Thus, in animated scientific movies based on drawing, many transformations can take advantage of this absence. The film "Birth of a Brain" (Movie 2016) can be considered as a demonstration of those possibilities: drawings are composed and recomposed, the spectator's gaze travels across the compositions, and important scale changes occur without inducing confusion or an excessive vertigo sensation. In practice, all those transformations are produced today in the context of "compositing" software. This word gives name to an important category of production tools, which offer the possibility of manipulating multi-layer images and altering those layers with special effects and transformations. Such software gives concrete access to the fundamental operations that we began to distinguish in a geometry without a point of view: animating a form along a trajectory, zooming inside a composition, and so on. Another important feature is the possibility of making layers appear or disappear, by controlling their opacity and the way their textures are mixed. Consequently, both movement *and* texture can have equal importance in the context of compositing software, if used properly.

In "Birth of a Brain", in addition to these basic operations, other more complex transformations have been used. At certain times, cells are moving or differentiating, so the movie has to show significant transformations of their structures. At other times, it is necessary to turn the embryo around and slightly change the view. In order to address such transformations related to changes in the aspect of living forms, we have two complementary possibilities. The first is to draw a sequence of drawings and to display it at the appropriate frame rate. This is the strategy in traditional animation, which is a natural extension of the sequence drawing that we described earlier. The second possibility is to compute intermediate images between drawings, using a technique called morphing. I will briefly describe these two options in the context of "Birth of a Brain", before analysing morphing in more detail.

"Birth of a Brain" uses traditional animation for a few very brief actions, for example when a microglial cell phagocytises a dead neuron. In this case, displaying several drawings from a sequence, one after the other, produces an obvious appearance of motion. However, an image sequence can be used with different timing. The display of the sequence of embryo drawings (Fig. 8.2b) is spanned throughout the movie, using cross-fades for transition between one step and another: here, we rely on the more long-term memory of the spectator to induce the concept of a slow, progressive transformation. In any case, fast or slow, the feeling of transformation induced by such image sequences cannot be entirely continuous or fluid: if displayed at a slow rate, one feels each different step, and if displayed fast, one feels a vibration of strokes. This vibration, by the way, is not a defect: it can give a sense of energy to animated drawings.

However, in certain cases, transforming the drawings in a continuous way, without vibration, can prove very useful. This is one of the main purposes of morphing-based animation. With this technique, an interpolation between two consecutive drawings in a sequence can be computed, in order to produce a fluid transition. This opens up new possibilities for guiding the attention of the audience along a scientific narrative. For example, in the early stages of embryo development, in "Birth of a Brain", the body of stem cells moves up and down while they divide. With conventional animation techniques, such repeated movement would cause a lot of visual distraction, and make the narration difficult. Morphing animation helps by creating a fluid movement for a whole population of cells, without over-burdening the attention of the audience.

Morphing can also be used to create transitions between two views of a body. In that case, the resulting animation is a turning movement. This operation can be very important in scientific transmission, since the most efficient view for exploring an object is often slightly different to the most efficient view for explaining a more detailed structure or process related to it. Providing such a transition between two views to an audience can be critical when explaining the early stages of the embryo, because most spectators will not know where to direct their attention. In that case, a slight turning movement, coordinated with pan and zoom, can be essential in order to orient properly the attention of the viewer. The fluidity of morphing is very important here, because it drives the attention far more efficiently than a simple cut transition between two images, due to the optic flow induced by continuous texture transformations.

8.4.4 Inside Morphing: Warping and Cross-Fade

Technically speaking, morphing appears to be a way of calculating intermediate images between two images. In practice, it breaks down into two operations: the first is an image distortion, called "warping" and the other is a crossfade. Interestingly, warping was invented in the 1980s during the rise of digital animation (Heckbert 1989) while crossfade was already being used by Daguerre in 1822. Before he invented his photography technique, he had the idea to paint the two sides of a canvas and to mix both images by illuminating the back. Strictly speaking, morphing cannot be reduced to either old or new technology: it uses an interesting combination of the two.

In order to understand how warping and cross-fade are combined, consider for example (Fig. 8.6a) two drawings of a mermaid inspired by a vintage merry-go-round, seen from a slightly different angles. From these two views, we will create a slow turning drawing.[2] First, using appropriate software, we will define deformation constraints on each of the two drawings, in the form of Bézier curves as shown in colour (Fig. 8.6b). Each of these curves is two-dimensional, so the curves on one drawing include no information about depth a priori. However, those curves can be animated: we thus specify how the curve associated to the left cheek in drawing A must be transformed to go to land on the left cheek in drawing B. While this transformation occurs in the plane, it integrates some information about space, since the transformation of the curve, while we turn the object around, depends on the depth. The set of all these curves allows us to define a global planar transformation of the image (Fig. 8.6c) which brings all the aspects represented in the drawing A to the place which corresponds to them (more or less precisely) in drawing B. We thus obtain two sequences of images: one which distorts drawing A in order to put it in the position of drawing B, and the other doing the opposite (Fig. 8.6d, e).

However, half-way through the respective transformation (warping) of A and B, the textures of the strokes begin to lose their adaptation to the space they should suggest. To compensate for this problem, we have to make a crossfade between the two sequences: while A is deformed, its texture is gradually replaced by that of B, so that at the end of the transformation the strokes of B have taken the place of those of A, with the correct texture for this angle of view. The whole of this transformation induces the sensation that the drawing turns like a rigid object, according to a vertical axis (Fig. 8.6f; see online resource Chabrier (2014) for an animated visualization).

---

[2]This merry-go-round can be seen in the Musée des Arts Forains in Paris. In the late nineteenth century, adults could take a ride on it for the simple pleasure of feeling the world turn slowly.

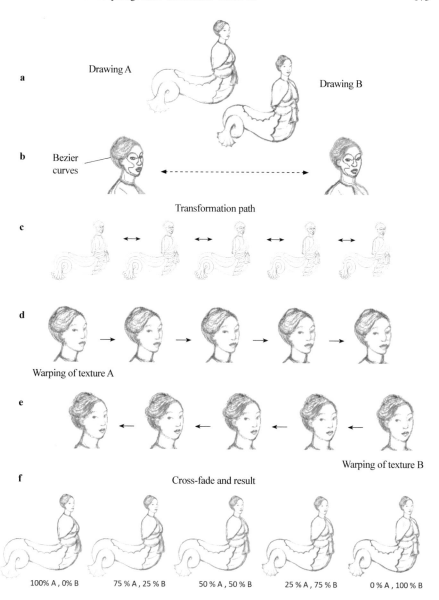

Fig. 8.6 Step-by-step description of morphing between two drawings: in this case, Bezier curves allow fine control of the transformation of strokes. **a** A draftsman creates 2 drawings. **b** The animator specifies corresponding curves on the 2 drawings. **c** A transformation path is computed on the basis of interpolation between those curves. **d** The drawing A is transformed (warped). **e** The drawing B is transformed (warped). **f** Finally, a transformation from animated texture A to animated texture B is computed (Cross fade)

However, all mathematical operations have been performed in the two-dimensional plane, where ordinarily only rotations with an axis perpendicular to the plane are permitted. By introducing a well-controlled non-linear deformation (i.e. warping) *and* a change of texture (i.e. cross-fade), the morphing thus makes it possible to evoke tri-dimensional transformations of drawings.

8.4.5   What Does Morphing Enable Access to?

In the above example, I showed that morphing could produce a visualization of rigid movements in three dimensions, using two drawings. Keeping in mind that drawings are made of strokes, we can try to establish if the concept associated with stroke in China, as we saw at the end of the previous section, could be used in that situation. I, personally, am tempted to say that both "brush" and "stroke" are taken into account in that technical process, thanks to the association of curve movements and texture changes. This might indicate that morphing can be considered not only as a special effect, but may give access to fundamental aspects of space–time geometries in the brain. We will consider here the kind of imagination that morphing gives access to, and progressively turn to more geometric aspects of the control of deformation, which may have an interesting scientific significance.

To begin with, morphing gives access to metamorphosis. This concept has a long history, from poetry to science. In his famous book "On Growth and Forms" (1917), long before digital imaging existed, D'Arcy Thompson demonstrated the possibility of geometric transformations between the bodies of two species. Planar transformations between the side view of various fishes or animal skulls, visualized through a grid, allowed him to question evolution and development (Fig. 8.7). The scientific concept had limited scientific success, but the concept of transforming one image into another remained. As we just saw, it was realized in the 1980s with morphing, largely outside of any scientific endeavour. A typical example would be the film "Willow", in which the viewer could watch a tiger turning into a woman. Because of the difficulty of controlling transformations precisely, morphing and warping were usually associated with transformations defying physical reality: significantly, one of the first software packages specifically for creating these effects was named "Elastic Reality".

In addition, in line with the work that I was able to carry out between 2000 and 2016, morphing can be used to visualize and control other kinds of changes in the form and structure of a body: rotations, articulations, rigid movement, and of course deformation when necessary. Ultimately, according to my experience of film-making, a multi-layered approach in morphing appears to be the most complex transformation in the compositing repertoire. Without it, it is very difficult to find any coherence in the disparate set of so-called "2D" animation techniques. Conversely, since the concept of morphing is closely related to drawing, it allows us to better integrate hand-made drawings in the context of fluid digital animation, from paper to screen. This approach gives access not only to spectacular metamorphosis, but to a subtler

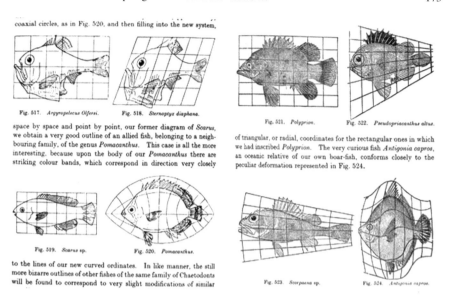

coaxial circles, as in Fig. 520, and then filling into the new system,

Fig. 517. *Argyropelecus Olfersi.*   Fig. 518. *Sternoptyx diaphana.*

space by space and point by point, our former diagram of *Scarus*, we obtain a very good outline of an allied fish, belonging to a neighbouring family, of the genus *Pomacanthus*. This case is all the more interesting, because upon the body of our *Pomacanthus* there are striking colour bands, which correspond in direction very closely

Fig. 521. *Polyprion.*   Fig. 522. *Pseudopriacanthus altus.*

of triangular, or radial, coordinates for the rectangular ones in which we had inscribed *Polyprion*. The very curious fish *Antigonia capros*, an oceanic relative of our own boar-fish, conforms closely to the peculiar deformation represented in Fig. 524.

Fig. 519. *Scarus sp.*   Fig. 520. *Pomacanthus.*

to the lines of our new curved ordinates. In like manner, the still more bizarre outlines of other fishes of the same family of Chaetodonts will be found to correspond to very slight modifications of similar

Fig. 523. *Scorpaena sp.*   Fig. 524. *Antigonia capros.*

Fig. 8.7 The principle of morphing anticipated by D'Arcy Thompson in his book «On growth and form» (1917). In practice, controlling morphing with such a grid allows insufficient control of the transformation of strokes

kind of imagination, based on the propensity of the drawings to be transformed by the imagination of their viewers.

Before the advent of morphing, there was no standard technique for visualizing what we imagine when we assess the possible transformations of a drawing. Depending on the animation approach, either the movement or the textures had to be greatly simplified, despite the fact that both contribute strongly to the spatiality of the stroke. With the use of morphing, during the 2010s, it was possible for me to animate Leonardo da Vinci's drawings (Movie 2013), as well as the drawings from the Chauvet Cave (Movie 2011–2016), while respecting the texture and geometry of the original strokes. These animations are not intended to develop a story that goes beyond the intention of the original artists. They simply seek to make visible a propensity of the drawing to be transformed in a certain way. But for this to happen, the animation process must not distort the geometry of the drawing. In particular, strokes with a spatial value must not be treated as lines or areas in a flat frontal plane.

A remarkable aspect of morphing is that it forces the animator to become much more aware of the transformations he uses. Indeed, rotations in three dimensions raise perceptual problems, depending on the way they are projected on a plane: a rotational movement whose axis is *in the plane* of the image will result in a linear trajectory of textural elements, as in the case of the mermaid. Alternatively, a rotational movement whose axis is *perpendicular to the plane* will move the texture of the image along a circular trajectory around a point of rotation. Those two particular types of spatial rotations are quite easy to specify from inside the projection plane, but general cases are more puzzling. If we show a head rising from a three-quarter view, for instance, the axis of this rotation (in space) will be neither in the image plane nor perpendicular to it. In such cases, if we want to animate the texture of the image according to this rotation, it will be necessary to specify a combination of rectilinear trajectories and circular trajectories. The morphing animation has to deal with this difficulty of specifying the appropriate transformations while staying in a plane. In practice, for example if we want to animate cave art like in Movie (2011–2016), specifying the transformation is often impossible without hand-made reference sketches on paper (Fig. 8.8).

In 3D synthetic images, this kind of problem does not arise: the animator has three parameters of rotation in space to control the movement of the objects he manipulates. Morphing therefore introduces a difficulty, due to the fact that it uses coordination between transformations "in the plane" and texture changes. Here we can observe that the brain largely uses a system in the form of maps, which appears more adapted to planar deformations than to tri-dimensional transformations. An appreciation of the specific difficulty of morphing could therefore strengthen the idea that within the brain, taking into account rotations in space requires a more heterogeneous process than what one might imagine.

Finally, we should bear in mind that the morphing of drawings does not use a camera, and consequently does not necessarily use a point of view. We have seen above that geometry without a point of view was based on the possible relations between the drawings, without any need to explicitly structure the space with a three-dimensional grid. Similarly, the morphing technique that I described is based on the use of curved lines to specify the transformation between drawings, without using a "deformation grid". It should be noted that a morphing action can be specified with an explicit grid (as in the images of D'Arcy Thompson), but in practice this approach turns out to be far too imprecise. Consequently, the effectiveness of Bézier curved lines in specifying the transformation of the main strokes in a drawing might make sense. As Flash remarked, Bézier curves are parabola, and such parabola seem to be used by the brain to generate tracing movements, because of their curvature properties in equi-affine geometry. Thus, morphing animation could have an epistemological role in addition to its applications in film making: it could also highlight the role of curvature in the representation of space in the brain.

8.5   Redefining the Epistemological Role of Drawing: A Work in Progress

Since drawing began to be replaced by photography, its epistemological role has not been clear. The digital revolution introduced even more tools, seemingly enabling the same tasks of representation as drawing, but without the constraints of materiality or lengthy practical training. But during the last decade, through a number of publications, workshops, movies and exhibitions, it has become evident that drawing gives access to a form of knowledge, or concentration, that could not be achieved through

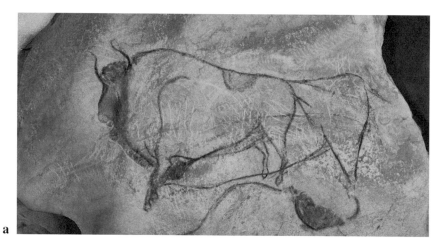

a

b

Fig. 8.8 Raw material for an animation of cave drawings with warping and morphing. **a** The (planar) texture of the image can be transformed in such a manner that the animation remains coherent with the spatiality of the representation. **b** Working on the computer is not sufficient: new drawings on paper, with real strokes, are often mandatory to specify this transformation correctly. Note that this association of three figures generates a composition without a point of view. Photography Marc Azéma, drawing R. Chabrier

other media. In this new context where animated images are at least as important as static ones, redefining the epistemological role of drawing seems necessary.

Such a task can be expressed more simply: what does drawing give access to? In this text, I focused on how drawing could provide access to various geometries of living beings. As I have demonstrated, thinking in terms of transformations, thanks to the model provided by modern geometry, makes it easier to compare a number of drawing practices, with different historical and epistemological backgrounds. The concept of stroke seems a mandatory means to unify them. But if we want to think and investigate further, in order to grant to the concept of stroke the importance it deserves, differential geometry as well as Chinese painting treatises appear to be required. Acting as a bridge between such distant domains, in terms of space, time and logic, might be one of the major epistemological roles of drawing.

This role is more than that of an academic pastime, since today the representation of life raises important issues. As far as geometry is concerned, we currently live in a paradox: while projective geometry is no longer taught to anyone apart from specialists in mathematics and computer graphics, central perspective and flat projection have never been so present in the images and films that are available to the public. By reducing the world of representations to a dialectic between "2D" and "3D", it is likely that we have lost in the sciences, arts and architecture, an important aspect of our ability to take into account the spatiality and diversity of living beings in our representations. Since the aspects of life that are not properly represented tend to be neglected in a world dominated by images, this may have contributed to a serious lack of awareness of environmental problems in recent decades. On this basis, emphasising a geometry without a point of view, where curvature plays a major role, seems necessary if we want to re-develop the ability to represent the spatiality of life in an accessible way.

Designing appropriate training that integrates the practice of drawing might be the beginning of a solution for re-developing access to this spatiality. For three years now, I have been running an experimental course for engineering students at the Ecole Polytechnique (Palaiseau) entitled "Drawing and Animation for Scientific Transmission", where the basics of representation are addressed simultaneously through drawing and animation. Short courses for researchers in cell biology and biophysics have also been successfully tested at Institut Curie, as well as a re-introduction of drawing by hand at different levels of scientific transmission (Chabrier and Janke 2017). Similarly, the scientific iconography department of the National Museum of Natural History recently established, under the direction of Didier Geffard-Kuriyama, a doctoral module in scientific drawing that had not existed for twenty years. Thus an interesting dynamic seems to be emerging in France in the scientific field. In addition, interest in drawing today cuts across the arts as well as design, architecture, medicine and other disciplines. In the United States and the UK, the Thinking through Drawing network, initiated by Kantrowitz et al. (2011), leads meetings, workshops and interdisciplinary debates that testify to the vivacity of the field.

As far as I know, the way in which a stroke produces a sensation of space, and the way in which several strokes cooperate to construct a spatial representation has not yet been modelled more widely. Today, several approaches seem to be converging towards a better understanding of this phenomenon using scientific methods: one can quote again the works of Viviani and Flash (1995), which introduced a power law in the neuronal planning of trajectories by connecting curvature and tangential velocity, and the subsequent advances when a more extensive connection with geometry was made (Flash and Handzel 2007; Bennequin et al. 2009). Creative robotic research currently aims at reproducing the process and the structure of the stroke in a more complete way, with a goal of developing drawing robots (Tresset and Fol Leymarie 2013; Gülzow et al. 2018). In this field, is worth noting that the team of Frederic Fol Leymarie at Goldsmiths, University of London works alongside on the modelization of calligraphy (Berio et al. 2017), and on life sciences problems involving 3D representations, such as the visualisation of folding and docking of proteins (Tod et al. 2015).

In general, modern scientific drawers have to use 3D information. Hence, even though they can be distant, the worlds of drawing and computer-assisted projective geometry (i.e. 3D), are no longer separated: in fact, they are complementary, and a large amount of time can be devoted to mixing the two types of information. In computer graphics, attempts to create a more direct link between planar representation and 3D also exist. The "sketch-based modelling" approach consists of interpreting as 3D information diagrams or drawings made in two dimensions. Several specific cases have been approached in this way, for example the spatial interpretation of blood vessel schematics (Palombi et al. 2011), notably by postulating that ellipses can be interpreted as sections of vessels. Currently, at Ecole Polytechnique, I am working with Marie-Paule Cani's team on a broader spatial interpretation, from strokes on paper to their display on an autostereoscopic screen.

Given these activities in various complementary fields, a "true" modelization of strokes may appear as possible in short term. Indeed, modelling a phenomenon is often considered a positive step for scientists. In artistic practice, however, nothing is more dangerous than thinking about the stroke through modelization. In my experience, the drawer always runs the risk of focusing his attention on an aspect of the stroke that could have been important yesterday, but which is no longer relevant today. If we want to develop robust research on the spatio-temporality of the stroke, in terms of transformations, we must therefore rely on a safeguard that prevents the concept from closing up and becoming sterile. The mere practice of drawing in real life, on real paper, provides such a safeguard. But once again, it is a Chinese concept which seems to best account for how the world can be connected to the stroke (and how the wealth of associated geometries in the brain can harmonize, if we can risk using such modern ideas). This is the "Oneness of brush stroke", also translated as "Unique brush stroke", expressed by the painter Shitao at the beginning of his treatise (1710, translation by Coleman 1978):

> The principle of oneness of strokes is such that from non-method method originates; from one method, all methods harmonize [...] With regards to the delicate arrangement of mountains, streams, and human figures, or the natural characteristics of birds, animals, grass, and trees, or the proportions of ponds, pavilions, towers, and terraces, if one's mind cannot deeply penetrate into their reality and subtly express their appearance, one has not yet understood the fundamental meaning of the oneness of strokes.

With or without a brush, the whole treatise of Shitao can still be read today as an invitation to access the diversity and the dynamics of the world through the use of strokes. Although this citation should not be interpreted without its historic and Taoist context, it seems a rather clear warning of the risks of reducing the stroke too quickly to one or other of its applications. Methods and rules can be useful, but non-method is really at the core of drawing, especially in the daily use of a sketchbook (Fig. 8.9).

8.6 Conclusion

The main goal of this text was to introduce a new, more effective and relevant vocabulary, in order to understand the epistemological role of drawing. Thanks to the work done by mathematicians, the word "transformation" now signifies an extremely powerful and general concept that covers events related to both movements and changes. If we use this concept of transformation in the context of drawing, we find that it is closely associated to the concept of "stroke".

"Stroke" provides an excellent alternative to the word "line" in many situations. Thanks to its use in the field of Chinese painting, it embraces the materiality, the spatiality and the temporality that develop in drawing, and especially in the sketch. Using the word "line" often results in implicitly reducing the stroke to a graphic object in two dimensions. Where possible, I suggest reserving "line" and "curve" for more explicitly abstract and/or mathematical uses, since in such contexts those concepts are far more clearly defined.

Finally, the word "texture" puts a necessary emphasis on some of the dimensions of the stroke that are currently the most neglected because of the dematerialization of tools for design and creativity. Animation approaches such as morphing can be used to make the role of textures in the representation of space more evident.

Thus, using more adapted concepts, we have seen that it is possible to shed light on a geometry without point of view, which has not previously been fully described. It appears to have constituted a fundamental basis for the drawing of living beings since the very origins of this practice. The characteristics of this geometry in terms of composition, recomposition and movements also allows us to understand better how to develop the potential of modern drawing approaches, for scientific or non-scientific purpose. In animation, museography or design, an important challenge today is to complete and balance the extensive development of tools and images based on flat projection and central perspective with a geometry that does not rely on a unique point of view. This geometry is easy to access thanks to drawing, and its maintenance may be equally important for the human brain. Its re-development might create the opportunity for a new association between mathematics, drawing and neurosciences.

Fig. 8.9 The epistemological role of drawing appears to be related to the "stroke". This concept has been developed primarily in Chinese civilization, on the basis of painting (or drawing) with brush and ink. The stroke was noticeably used for painting landscapes which do not use a point of view. Shitao and Lake Cao (1695)

Acknowledgements This work received support from PSL Research University and from the grants ANR-11-LABX-0038 and ANR-10-IDEX-0001-02.

I would like to thank Tamar Flash, Alain Berthoz and Daniel Bennequin for their kind invitation and their patience.

I also thank Emma Neave and Seymour Simmons for their help with writing in English.

Finally, thank you to the models of the life drawing workshop of La Grande Chaumière and to the many people involved in producing the films and morphing animations mentioned in the text.

Methods The morphing animations were mainly created with the Combustion compositing software and the associated Re:Flex morphing plugin.

References

Bennequin, D., Fuchs, R., Berthoz, A., Flash, T. (2009). Movement timing and invariance arise from several geometries. *PLoS Computational Biology*

Bennequin, D., Berthoz, A. (2017). Several geometries for movement generation. In J. P. Laumond, N. Mansard, J. B. Lasserre (Eds.), *Geometric and numerical foundations of movements*, STAR Series 117. Springer, Heidelberg

Berio, D., Calinon, S., Fol Leymarie, F. (2017). *Generating calligraphic trajectories with model predictive control*. In Graphics Interface 2017, Edmonton, Canada

Berthoz, A. (2009). *La simplexité*. Paris: Odile Jacob.

Bourdieu, P. (2013). *Manet, une révolution symbolique*. Seuil, Paris: Raisons d'agir.

Chabrier, R. (2016). Pouvoir de représentation contre pouvoir absolu : le dessin scientifique confronté à la perspective. In Mariannick Guennec, Véfa Lucas (eds) Les images du pouvoir, Univ.Brest, HCTI/UBS, Lorient

Chabrier, R., & Janke, C. (2017). The comeback of hand drawing in modern life sciences. *Nature Review Molecular Cell Biology, 19,* 137–138.

Coleman, E. J. (1978). *Philosophy of painting by Shih-T'Ao*. The Hague: Mouton.

Thompson, D. (1917). *On growth and form*. Reedition

Daston, L., & Galison, P. (2007). *Objectivity*. New York: Zone Books.

Flash, T., & Handzel, A. (2007). Affine differential geometry analysis of human arm movements. *Biological Cybernetics, 96,* 577–601.

Fong, W. C. (2003). Why Chinese painting is history. *The Art Bulletin, 85*(2), 258–280.

Fraipont, G. (1897). *L'art d'utiliser ses connaissances en dessin*. Paris: H. Laurens.

Grossos, P. (2017). *Signe et forme: Philosophie de l'art et art paléolithique*. Paris: Les Editions du Cerf.

Gülzow, J., Grayver, L., & Deussen, O. (2018). Self-Improving Robotic Brushstroke Replication. *Arts, 7,* 84.

Heckbert, P. S. (1989). Fundamentals of texture mapping and image warping. Department of Electrical Engineering and Computer Science, University of California

Heider, F., & Simmel, M. (1944). An experimental study of apparent behavior. *The American Journal of Psychology, 57*(2), 243–259.

Ivey, T., & Landsberg, J. M. (2003). *Cartan for beginners: Differential geometry via moving frames and exterior differential systems*. Providence: American Mathematical Society.

Jullien, F. (2003). La grande image n'a pas de forme, ou du non-objet dans la peinture, Seuil, Paris. English edition (2009). The great image has no form, or on the nonobject through painting (Trad: J. M. Todd). University of Chicago Press

Jullien, F. (2009). Les transformations silencieuses. Grasset, Paris. English edition (2011). The silent Transformations (Trad: K. Fijalkowski, M. Richardson). Seagull Books, Calcutta

Kantrowitz, A., Brew, A., & Fava, M. (Eds.). (2011). *Thinking through drawing: Practice into knowledge*. New York: Columbia University, Teachers College.

Korzybski, A. (1933). *Science and sanity: An introduction to non-aristotelian systems and general semantics* (5th edn.). Institute of General Semantics (1994)

Klein F (1872) Vergleichende Betrachtungen über neuerer geometrische Forschungen. Deichert

Leroy-Gourhan, A. (1965). *Préhistoire de l'art occidental*. Mazenod, Paris

MacLaren N (1960) Vertical lines (short movie). Office National du Film Canadien

Nicolaides, K. (1941). *The natural way to draw*. Boston: Houghton Mifflin.

Palombi, O., Pihuit, A., Cani, M. P. (2011). 3D modeling of branching vessels from anatomical sketches: towards a new interactive teaching of anatomy. In *Proceedings of surgical and radiologic anatomy*

Panofsky E (1927) Perspective as a symbolic form. Reedition Zone Books (1991) New York

Popescu-Pampu, P. (2018). De la topographie à la géométrie. In P. Picouet (Ed.), La carte invente le monde. Presses Universitaires du Septentrion, Villeneuve d'Asq

Simmons, S. (2011). Philosophical dimensions of drawing instructions. In A. Kantrowitz, A. Brew, M. Fava (Eds.), *Thinking through drawing : Practice into knowledge*. Teachers College, Columbia University, New York

Shitao (1710) Les propos sur la peinture du moine Citrouille-amère. Edition française : Hermann (1984) (Trad: P. Ryckmans). Paris

Todd, S. et al. (2015). FoldSynth: interactive 2D/3D visualisation platform for molecular strands. In: *Proceedings of the Eurographics workshop on visual computing for biology and medicine (VCBM)* (pp. 41–50).

Tresset, P., Fol Leymarie, F. (2013). Portrait drawing by Paul the robot. *Computers and Graphics, 37*(5)

Viviani, P., & Flash, T. (1995). Minimum-jerk, two-thirds power law, and isochrony: Converging approaches to movement planning. *Journal of Experimental Psychology: Human Perception and Performance, 21,* 32–53.

White, S. J., Coniston, D., Rogers, R., & Frith, U. (2011). Developing the Frith-Happé animations: A quick and objective test of theory of mind for adults with autism. *Autism Research, 4,* 149–154.

Movie References

Birth of a Brain (2016). https://vimeo.com/183004485

Leonardo da Vinci—Motion Picture (2013). https://vimeo.com/84689530

Cave art animation studies (2011–2016). https://vimeo.com/171478742

Online Resource

Chabrier, R. (2014). Le Morphing, l'art de transformer les images. https://images.math.cnrs.fr/Le-Morphing.html

Part III
The Arts: Music

Chapter 9
Egocentric Dynamic Planar Organization of the Angular Movements of the Arm During a Violinist's Performance of a Mozart Symphony

G. Cheron, M. Petieau, A. M. Cebolla, C. Simar, and A. Leroy

Abstract We here studied the possibility that the coordination of the upper limb segments acting on the hand controlling the bow of the violin during a musical performance could be organized in a planar covariation pattern. Two music masters from the Queen Elisabeth Music Chapel of Belgium played the cadence of the first movement of Mozart's concert symphony for violin and viola face to face, both equipped with optoelectronic markers for 3D movement, EMG, and EEG. We demonstrated that the best covariation plane was obtained when the flexion-extension angle of the elbow (α), wrist (β), and the angle formed by the fingers and the bow (Y) are dynamically projected into the triangular plane formed by the left upper limb and violin. This planar representational geometry only emerges in the actual situation when the movement of the body was indirectly taken into account by projecting onto the egocentric dynamic plane and not on the allocentric passive plane.

9.1 Introduction

With only this sentence, *"We listen to the music with our muscles,"* Friederich Nietzsche (cited in Sacks 2007) reignited interest in the deep interaction between music and movement, opening the door to basic questioning about the production of music by the brain.

G. Cheron (✉) · M. Petieau · A. M. Cebolla · A. Leroy
Laboratory of Neurophysiology and Movement Biomechanics, Université Libre de Bruxelles, Brussels, Belgium
e-mail: gcheron@ulb.ac.be

G. Cheron
Laboratory of Electrophysiology, Université de Mons-Hainaut, Mons, Belgium

C. Simar
Machine Learning Group, Université Libre de Bruxelles, Brussels, Belgium

A. Leroy
Haute Ecole Provinciale du, Hainaut-Condorcet, Mons, Belgium

© Springer Nature Switzerland AG 2021
T. Flash and A. Berthoz (eds.), *Space-Time Geometries for Motion and Perception in the Brain and the Arts*, Lecture Notes in Morphogenesis,
https://doi.org/10.1007/978-3-030-57227-3_9

As with any type of sound, music results from compression-decompression cycles of ambient air induced by a movement by the non-living (e.g., the sound of the wind or the sea waves) or living entities in the environment. In this latter case, the movement can be highly organized by the brain in such a way that it becomes psychologically significant, and may induce pleasure (Zatorre and Salimpoor 2013; Gold et al. 2019). When music is produced, it immediately creates a specific mental state not only in the musician (the producer), but also in the listener (the receiver), spontaneously entering a complex loop of resonance that may be conducted into motor action, such as dance. The universal treat in humans of moving to the music was recently reviewed by Levitin et al. (2018), who highlighted the importance of movement, rhythm, and synchronization in final perception and related cognition.

The recruitment of brain motor areas, such as the supplementary motor area and the cerebellum, during musical perception (Chen et al. 2008; Palomar-García et al. 2017) reinforces action-perception processing into a rhythmic template. To reach such rhythmicity, musicians must acquire perfect coordination of the different limb movements implicated in the musical gesture through learning and practice involving movement repetition. This skilled practice induces brain plasticity (Steele and Zatorre 2018) and leads to highly stereotyped movements in which a final graceful personal touch of creativity could produce a masterpiece.

In this context, musicians and sport performers are in demand of a particular sensation related to their performance, described as the 'flow', a concept from positive psychology (Csikszentmihalyi 1975) for which new perspectives from neuroscience have been proposed (Cheron 2016). The fact that the flow occurs during the top performance in music implies that at least three main elements combine to achieve this state of consciousness: (1) optimization of the descending motor commands, (2) central appropriate mental resting state, including memorized and predictive items, and (3) ascending pluri-modal sensations closing the loop between action and sensation.

The activity patterns of multiple muscles captured by electromyography (EMG) may reveal the basic motor coordination dynamics of the final gesture (Scholz & Kelso 1990; Kelso 1995). In accordance with Bernstein's view (Bernstein 1967), dynamic patterns emerge through exploration of available solutions to the redundancy problem, leading to the selection of preferred movement strategies and organizational principles.

A new perspective is opening the field of biomechanics to brain oscillatory models (Cheron 2015), which is in line with the inside-out concept recently promoted by Buzsáki (2019). As explained previously by Berthoz (1997), the brain is not a passive entity, but a biological organ selected in order to produce motor actions by which it actively interrogates and predicts the environmental variables. Although the highly complex movement realized for music production is probably one of the most intellectual, emotional, and creative human actions, it probably shares the same organizational principle as a more common action, such as locomotion.

For example, in toddlers, the rhythmic leg patterns of walking progressively emerge through repeated cycles of action and perception ending in a planar covariation pattern (Cheron et al. 2001). This intersegmental coordination rule previously characterized in adults (Borghese et al. 1996; Lacquaniti et al. 1999; Hicheur et al. 2006; Barliya et al. 2013; Ivanenko et al. 2008) was modeled by simple oscillators coupled with appropriate time shifts (Barliya et al. 2009), indicating that a combination of oscillatory commands may be used for movement coordination (Hoellinger et al. 2013). The contribution of hardware structures, such as reciprocal inhibitory connections of the central pattern generator (CPG), may directly control the motions of the different limb segments by encoding the harmonics of the elevation angles (Lacquaniti et al. 2002).

By analogy, we studied the possibility that the coordination of the upper limb segments acting on the hand controlling the bow of the violin during a musical performance could also be organized in a planar covariation pattern. Obviously, the violinist's movement is entirely different from and more kinematically complex than locomotion. In this latter case, the walking rhythmicity is oriented in a forward direction and the body equilibrium obeys the inverse pendulum law, in which the gravitational force plays an important role (Ivanenko et al. 2004). During the violinist's movement, the global posture of the body rests relatively in place except for the "musical" oscillations of the trunk in the standing or sitting posture, which probably facilitate the postural control of the whole body when challenged by other biomechanical and instrumental constraints, making the organizational principle of the skilled gesture more complex.

9.2 Results and Discussion

In the present chapter, we describe a preliminary experiment performed by two participants who are music masters from the Queen Elisabeth Music Chapel of Belgium. These masters played the cadence of the first movement of Mozart's concert symphony for violin and viola face to face, both equipped with optoelectronic markers for 3D movement, EMG, and EEG. We focused on only the 3D kinematics of one of the musicians in order to decipher the possible representational geometry of this motor behavior. The musicians performed a series of 10 identical pieces, of 80 s duration each, of the first movement of the symphony. At the end of each execution, they gave themselves a performance score ranging from 0 to 10.

The complexity of the violinist's movements was represented by the 3D movement of the distal point of the bow during the musical performance (Fig. 9.1a). This complex trajectory results from the combination of movements of the right upper limb acting on the bow, the left upper limb maintaining the violin, and those of the body. In the present study, the small movements of the violinist's finger tips acting on the strings were not taken into account, and we focused our analysis on the right upper limb mainly responsible for the bow movements. The bow trajectory was predominated by the vertical reversal displacement (in the Z direction) of approximately 40 cm, with smaller displacements in the Y and X directions. These movements were irregular and varied from 1 to 2.5 Hz (Fig. 9.1b). Figure 9.1c illustrates the

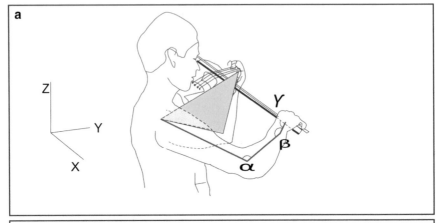

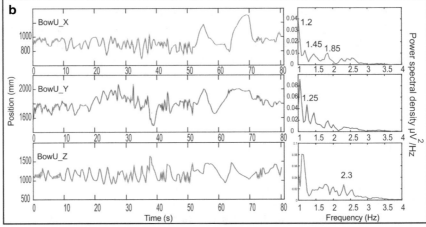

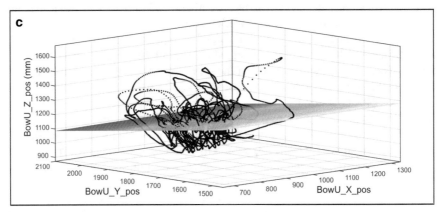

◀**Fig. 9.1** **a** Schematic representation of the 'dynamic plane' (yellow triangle) formed by the violin
body and the left upper limb. The three angles of the right arm (elbow (α), wrist (β), and the one
formed by the fingers and the bow (Υ)) are projected into the 'dynamic plane' or horizontal plane.
b The X, Y, and Z positions in the space of the upper extremity of the bow during the whole duration
of the musical performance is represented on the left side from top to bottom. The corresponding
fast Fourier transform (FFT) is presented on the right. **c** 3D representation of the movement of the
bow. A best fitting plane is represented for illustrative purposes

absence of any planar organization of the bow movement, and the best fitting plane
cuts the bow trajectories in the middle of the data points.

We aimed to test the possibility that the best covariation plane would be obtained
if the flexion-extension angle of the elbow (α), wrist (β), and the angle formed by the
fingers and the bow (Υ) are dynamically projected into the triangular plane formed by
the left upper limb and violin. Because this plane is moving as the musician played the
sonata, we named it the 'dynamic plane', in contrast to the fixed horizontal or vertical
planes. This idea is based on the fact that at least three functional constraints must be
taken into account for violin practice: (1) the postural control of the violin body in the
appropriate plane in order to insure the skilled bowing movement relative to the string
and the string crossings, (2) the complex reversal movement of the bow on the violin,
and (3) the global oscillating movement of the trunk related to the emotional content
of the music produced (Konczak et al. 2009; Rodger et al. 2012; Schoonderwaldt and
Altenmüller 2014). To compare the emergence of a possible plane when the projection
was made on the dynamic plane (egocentric) or another referential (allocentric), we
projected these three angles (α, β, Υ) into the horizontal plane. As the dynamic plane
formed by the left upper limb and the violin moved differently than the right upper
limb acting on the bow and resulted from varied shoulder abduction-adduction and
head-trunk movements, we illustrated the movement of this plane in reference to the
3D axis (X, Y, Z) (Fig. 9.2a–c, left). The amplitude of the plane movement concerned
all three axes with the same extent of 30–50°. The movements were irregular and
presented four FFT peaks at 1.0, 1.17–1.2, 1.35, and 1.55 Hz (Fig. 9.2a–c, right).

Figure 9.3a illustrates the temporal evolution of the three angles (elbow, α; wrist,
β; and the fingers, Υ) during the whole music piece, which was approximately
80 s. Notably, the FFT peaks of these angular evolutions were not exactly the same
as those of the dynamic plane; the slowest components at 1.0 and 1.17–1.2 Hz
were present, but higher peaks occurred at 1.85–2.0 Hz (Fig. 9.3b), indicating a
relative independence of the movements of the arm controlling the violin forming
the dynamic plane and the angular variation of the form-arm acting on the bow. These
angular values were then projected on the dynamic plane, forming a tridimensional
representation of a complex trajectory fitted on a best plane, with $R^2 = 0.94$ in
the illustrated example (Fig. 9.3c), a performance that was reproduced for all 10
pieces (mean $R^2 = 0.91 \pm 0.02$). In contrast, when the same angles were projected
on the horizontal plane, different angular profiles were obtained (Fig. 9.4a) with
different FFT profiles in favor of higher frequency peaks (Fig. 9.4b). The best fit was
significantly worse regardless of the repeated movement (mean $R^2 = 0.44 \pm .0.09$).

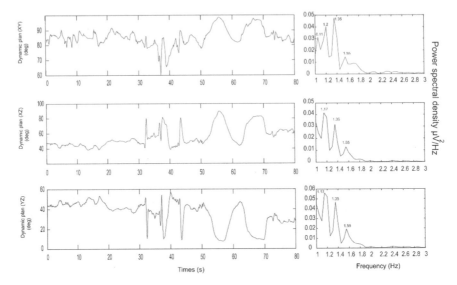

Fig. 9.2 Temporal evolution of the dynamic plane movement in X, Y, Z coordination during the musical performance (left) and the respective power spectral density resulting from FFT analysis (right)

This result was reproduced for the 10 executed pieces. Although the best performance score (recording 7) corresponded with one of the best correlation coefficients (0.95), and the worse score (recording 4) with the lowest correlation coefficient (0.87), these small differences (Fig. 9.5b) prevented us from testing the hypothesis that the 'flow' state was accompanied by the best planar correlation.

In addition, the orientations of the best fitting dynamic planes were well-conserved during these different performances, whereas the orientations of the planes obtained from the projection into the horizontal plane were not conserved (Fig. 9.5). This strongly indicates that the planar covariation resulting from the projection of the three angles of the form-arm acting on the bow on the dynamics plane formed by the other upper limb and the violin body could be considered a representational geometry of this complex movement. Representational geometry is understood here as a mathematical characterization of the inherent dynamic of the action-perception cycle (Kriegeskorte and Kievit 2013).

This planar geometry only emerges in the actual situation when the movement of the body was indirectly taken into account by projecting onto the egocentric dynamic plane and not on the allocentric passive plane. This finding may be due to the inherent coordination between the two form-arms and whole body movement.

Is this plane a logical and trivial consequence of a biomechanical coupling of the violinist's movement or a physiological indicator of neural control resulting from a simplified geometrical rule? The fact that the violinist reproduced the same piece 10 times presenting the same coordination plane (egocentric reference) with an R^2 close to 0.9 and the same conserved orientation of the normal vector. In contrast, in the

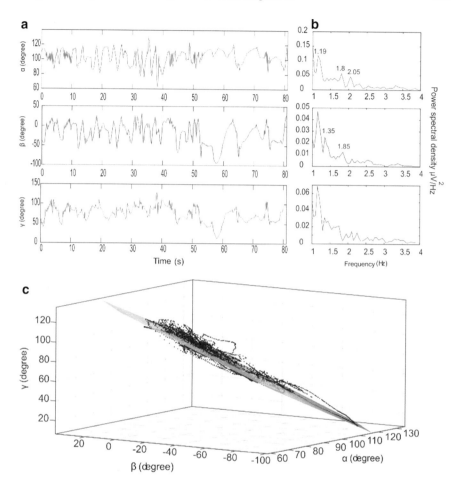

Fig. 9.3 **a** Temporal evolution of the three angles α, β, and Υ projected into the 'dynamic plane' from top to bottom during the sonata execution (left). **b** Respective power spectral density for FFT analysis. **c** The best fitting plane of these three angles with $R^2 = 0.94$

coordination plane (allocentric reference), the R^2 was not so strong and the normal vectors were more divergent in the repetition of the same musical piece. This result is clearly in favor of neuronal processing (Fig. 9.5).

The present transformation forming a planar representation encourages us to pursue a deeper mathematical analysis of the type of geometry (Euclidian, equi-affine, and full affine) as performed in Bennequin et al. (2009). We also intend to extend the present analysis to a possible role in the violinist's inherent movement oscillations, ranging from 1 to 3.5 Hz in the present case, a frequency range largely compatible with the neuronal oscillations of the brain. As already suggested (Cheron 2015) and tested for in locomotion (Hoellinger et al. 2013), the fact that brain signals are by nature oscillatory may induce the emergence of a representational geometry

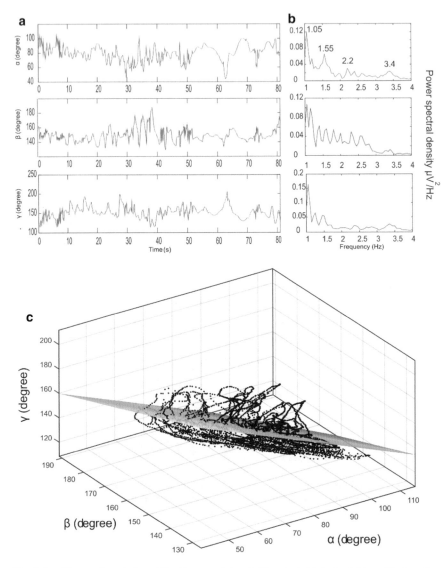

Fig. 9.4 **a** Temporal evolution of the three angles α, β, Ɣ projected into the horizontal plane from top to bottom during the sonata execution (left). **b** The respective power spectral density for FFT analysis. **c** The best fitting plane of these three angles with $R^2 = 0.44$

based on the combination of hierarchically organized oscillations (subcortical and cortical) acting on the brainstem and spinal "learning CPGs" (Yuste et al. 2005). These oscillatory commands could generate rhythmic movements that lead to the emergence of spontaneous dynamics. Another advantage of introducing a dynamic plan is that this combines the postural movement of the whole body acting on the stability of the musical instrument in one hand and the skilled control of the rhythmic

Fig. 9.5 **a** Comparative analysis of the normal vector of the best fitting planes resulting from the projection of α, β, and Ɣ into the 'dynamic plan' (blue vector lines) and the horizontal plane (red vector lines). **b** Histogram of the R square coefficient of the best fitting planes in these respective situations for the 10 performances

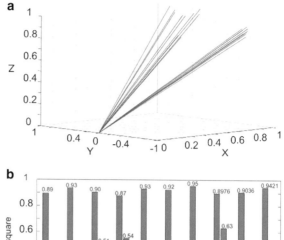

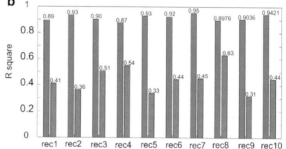

movement of the bow in the other hand. The present finding of an egocentric dynamic planar organization could represent an early first step in a long quest for the neuronal dynamics of musical arts.

Acknowledgements We would like to thank Miguel Da Silva and Augustin Dumay for their musical performances.

References

Berthoz, A. (1997). *Le Sens Du Mouvement*. Paris: Odile Jacob.

Barliya, A., Omlor, L., Giese, M. A., Berthoz, A., & Flash, T. (2013). Expression of emotion in the kinematics of locomotion. *Experimental Brain Research, 225,* 159–176.

Barliya, A., Omlor, L., Giese, M. A., & Flash, T. (2009). An analytical formulation of the law of intersegmental coordination during human locomotion. *Experimental Brain Research, 193,* 371–385.

Bennequin, D., Fuchs, R., Berthoz, A., & Flash, T. (2009). Movement timing and invariance arise from several geometries. *PLoS Computational Biology, 5,* e1000426.

Bernstein, N. (1967). *(1967). The Coordination and Regulation of Movements*. London: Pergamon.

Borghese, N. A., Bianchi, L., & Lacquaniti, F. (1996). Kinematic determinants of human locomotion. *The Journal of Physiology, 494* (Pt 3), 863–879.

Buzsáki, G. (2019). *The Brain from Inside Out*, Oxford Universiyt Press. edn.

Chen, J. L., Penhune, V. B., & Zatorre, R. J. (2008). Listening to musical rhythms recruits motor regions of the brain. *Cereb. Cortex N. Y. N, 18,* 2844–2854.

Cheron, G. (2015). From biomechanics to sport psychology: the current oscillatory approach. *Frontiers in Psychology, 6,* 1642.

Cheron, G. (2016). How to measure the psychological "Flow"? A neuroscience perspective. *Frontiers in Psychology, 7,* 1823.

Cheron, G., Bouillot, E., Dan, B., Bengoetxea, A., Draye, J. P., & Lacquaniti, F. (2001). Development of a kinematic coordination pattern in toddler locomotion: planar covariation. *Experimental Brain Research 137,* 455–466.

Csikszentmihalyi, M. (1975). *Beyond Boredom and Anxiety.* Jossey-Bass Publishers.

Gold, B.P., Pearce, M.T., Mas-Herrero, E., Dagher, A., & Zatorre, R.J. (2019) Predictability and uncertainty in the pleasure of music: a reward for learning?. *Journal of Neuroscience. Off. Journal of Social Neuroscience.*

Hicheur, H., Terekhov, A. V., & Berthoz, A. (2006). Intersegmental coordination during human locomotion: does planar covariation of elevation angles reflect central constraints? *Journal of Neurophysiology, 96,* 1406–1419.

Hoellinger, T., Petieau, M., Duvinage, M., Castermans, T., Seetharaman, K., Cebolla, A.-M., et al. (2013). Biological oscillations for learning walking coordination: Dynamic recurrent neural network functionally models physiological central pattern generator. *Frontiers in Computational Neuroscience, 7,* 70.

Ivanenko, Y. P., Dominici, N., Cappellini, G., Dan, B., Cheron, G., & Lacquaniti, F. (2004). Development of pendulum mechanism and kinematic coordination from the first unsupported steps in toddlers. *Journal of Experimental Biology, 207,* 3797–3810.

Ivanenko, Y. P., d'Avella, A., Poppele, R. E., & Lacquaniti, F. (2008). On the origin of planar covariation of elevation angles during human locomotion. *Journal of Neurophysiology, 99,* 1890–1898.

Kelso, J.A.S. (1995). *Dynamic Patterns: The Self-Organization of Brain and Behavior,* MIT Press, Cambridge, MA. edn.

Konczak, J., Vander Velden, H., & Jaeger, L. (2009). Learning to play the violin: motor control by freezing, not freeing degrees of freedom. *Journal of Motor Behavior, 41,* 243–252.

Kriegeskorte, N., & Kievit, R. A. (2013). Representational geometry: Integrating cognition, computation, and the brain. *Trends in cognitive sciences, 17,* 401–412.

Lacquaniti, F., Ivanenko, Y. P., & Zago, M. (2002). Kinematic control of walking. *Archives Italiennes de Biologie, 140,* 263–272.

Lacquaniti, F., Grasso, R., & Zago, M. (1999). Motor Patterns in Walking. *News Physiology Science International Journal Physiology Production Jointly International Union Physiology Science America Physiology Social, 14,* 168–174.

Levitin, D. J., Grahn, J. A., & London, J. (2018). The Psychology of Music: Rhythm and Movement. *Annual Review of Psychology, 69,* 51–75.

Palomar-García, M.-Á., Zatorre, R. J., Ventura-Campos, N., Bueichekú, E., & Ávila, C. (2017). Modulation of Functional Connectivity in Auditory-Motor Networks in Musicians Compared with Nonmusicians. *Cereb. Cortex N. Y. N, 27,* 2768–2778.

Rodger, M. W. M., Craig, C. M., & O'Modhrain, S. (2012). Expertise is perceived from both sound and body movement in musical performance. *Human Movement Science, 31,* 1137–1150.

Sacks, O .(2007).*Tales of Music and the Brain,* Vintage Book, Random House, NY. edn. New York.

Scholz, J. P., & Kelso, J. A. (1990). Intentional switching between patterns of bimanual coordination depends on the intrinsic dynamics of the patterns. *Journal of Motor Behavior, 22,* 98–124.

Schoonderwaldt, E., & Altenmüller, E. (2014). Coordination in fast repetitive violin-bowing patterns. *PLoS ONE, 9,* e106615.

Steele, C. J., & Zatorre, R. J. (2018). Practice makes plasticity. *Nature Neuroscience, 21,* 1645–1646.

Yuste, R., MacLean, J. N., Smith, J., & Lansner, A. (2005). The cortex as a central pattern generator. *Nature Reviews Neuroscience, 6,* 477–483.

Zatorre, R.J., & Salimpoor, V.N. (2013). From perception to pleasure: music and its neural substrates. *Proceedings of the National Academy of Sciences. U. S. A.,* **110**(Suppl 2), 10430–10437.

Chapter 10
Space, Time and Expression in Orchestral Conducting

Eitan Globerson, Tamar Flash, and Zohar Eitan

Abstract The art of conducting involves a formal set of gestures, designed to convey musical meaning through movement. Interestingly, this relatively simple collection of gestures enables the conductor to communicate highly complex musical messages to the ensemble, indicating a non-trivial interaction between movement kinematics and sound. The following book chapter discusses this phenomenon. We first introduce movement-to-sound cross-modal mapping, surveying behavioral and neurophysiological research suggesting how spatio-kinetic features may be "translated" into aspects of musical rhythm, loudness and pitch. This is followed by a detailed discussion of the kinematics of right-hand gestures in conducting and their musical meanings. The kinematic correlates of expressive conducting are discussed by introducing some of the basic principles governing human movement generation. These principles include the wish to maximize motion smoothness, captured by the "minimum jerk" model, and the isochrony principle governing motor timing. Notwithstanding the considerable work that has already been invested in analyzing music conducting, many of the secrets underlying musical conducting still remain to be unraveled. This can only be achieved through a multidisciplinary approach, involving a variety of research disciplines, such as social psychology, computational motor control, musicology, mechanics, acoustics, as well as cross-modal sound-to-movement mapping.

E. Globerson (✉)
Jerusalem Academy of Music and Dance, Jerusalem, Israel
e-mail: eitan.globerson@gmail.com

T. Flash
Weizmann Institute of Science, Rehovot, Israel

Z. Eitan
School of Music, Tel Aviv University, Tel Aviv, Israel

© Springer Nature Switzerland AG 2021
T. Flash and A. Berthoz (eds.), *Space-Time Geometries for Motion and Perception in the Brain and the Arts*, Lecture Notes in Morphogenesis,
https://doi.org/10.1007/978-3-030-57227-3_10

10.1 Why Conducting?

Conducting is an enigmatic profession. On the one hand, conducting gestures are relatively easy to learn. Anyone with intact motor control and bimanual coordination can master them in a rather short period of learning. Consequently, it is considered by many to be an easily acquired skill (Schuller 1999, p. 3). However, simple as it may seem, very few conductors have established major international careers, strongly suggesting that there is more to conducting than learning a set of movements. Conducting gestures are considered by experienced conductors to be basic tools, beyond which lies an entire world of unwritten secrets. These secrets are considered by some well-known conductors to be a heavenly gift, which cannot be formally taught. The late Pierre Boulez, for example, summarized his views of teaching conducting with the words: "*You can only teach how to start and stop*". Similarly, New York Philharmonic conductor, Anton Seidl (1850–1898) stated that: "*The ability to conduct is a gift of God with which few have been endowed in full measure. Those who possess it in abundance do not wish to write about it; for them the talent seems so natural a thing that they cannot see the need to discuss it. This is the kernel of the whole matter. If you have the divine gift within you, you can conduct; if you have not, you will never be able to acquire it. Those who have been endowed with the gift are conductors; the others are time-beaters*" (Seidl 1899, p. 215). Following this line of thought, textbooks of conducting only teach basic gestures (discussed later in this chapter), using more or less the same diagrams to illustrate them. Beyond that, one can find a great variability of approaches to the pedagogy of conducting, highlighting the controversial characteristics of the conducting profession, which has never been completely formalized (Scherchen 1989, p. 3–5). Conducting is a multi-faceted activity, requiring high-level musicianship, alongside excellent interpersonal communication skills, strong leadership abilities, verbal skills, motor coordination, emotional intelligence, and many more interdisciplinary qualities. This versatile combination of qualities is difficult, if not impossible to evaluate and quantify. It is as difficult to teach, since many of the required qualities of a good conductor involve personality characteristics, which are molded through life experience, rather than through formal training. Furthermore, conductors differ substantially in their implementation of the formal rules of the conducting gestures. Some use minimal movements, and in some cases, refrain completely from conducting. The famous video of Bernstein Conducting a Haydn Symphony using his eyebrows only, is a clear example of the difficulty in trying to formalize the art of conducting, considering the great variability in conducting methods, between and even within conductors. Growing trust between conductor and ensemble enables the conductor to be more and more minimal in the amount of information conveyed by formal conducting gestures, while dedicating more to the communication of general inspiration to the ensemble. Verbal instructions may also substitute for gestural information, not to mention the important role of facial expressions in communicating emotion to the ensemble. Hence, an evaluation of real-life conducting may require a weighted sum of multiple factors, some of them difficult, if not impossible to evaluate. This poses a theoretically unfeasible enterprise, which may lead to the conclusion that conducting is indeed impossible to evaluate in quantitative measures. This may explain the dearth of publications offering a quantitative evaluation of conducting gestures. However vague and enigmatic as it may seem, the art of conducting is taught at most university-level music institutes in the world, producing the next generation of professional conductors. The formal training of a conductor ubiquitously involves formal baton technique (i.e., right-hand gestures), as well as rehearsal planning, orchestral score reading, ear training, and other measurable musical skills. This set of tangible skills enables orchestras to perform with dozens of conductors every concert season, sometimes having no preparatory rehearsals before the performance. Hence, notwithstanding the extensive array of characteristics defining a good conductor, there seems to be a set of common traits characterizing a prototypic professional conductor. Otherwise, there would be no possible way for an orchestra to quickly adjust to the considerable turnover in conductors taking the podium in rehearsals and concerts. The most quantifiable attribute of conducting is the kinematics of hand movements. Right-hand movements in conducting are intended to convey rhythmic properties of the music being played, alongside articulation, loudness ("dynamics" in musical terms-we will use the term "loudness" to avoid confusion with the term dynamics used in the motor control literature), as well as emotional expression. Left-hand movements are dedicated mainly for communicating emotional expression, cues for specific sections or players, alongside additional information on loudness and articulation (McElheran 2004, pp. 37–38). Together, bimanual movements in conducting communicate a comprehensive musical statement to the musicians in the ensemble, portraying the highly complex emotional and esthetic information conveyed in music. Consequently, conducting offers an excellent field of study for anyone interested in the subtlety of motor movements, and their ability to convey a complex and personal message. Up-to-date studies of conducting gestures usually employ motion capture devices, alongside corresponding kinematic analysis algorithms, providing useful information on the kinematics of conducting movements. These have led to a greater understanding of the linkage between conducting gestures and the musical outcome. However, there is still a great amount of research work to be carried out, in order to achieve a greater understanding of the expressive component of conducting movements, and how they may lead to an inspiring performance. The current book chapter describes in detail the formal language of conducting, from kinematic and perceptual perspectives, employing a systematic review of musical parameters and their corresponding conducting gestures. The musical elements which will be addressed will include: beat, meter, loudness and articulation. All these parameters and their gestural representations in conducting have been the subject of various studies focusing on the language of conducting. More abstract phenomena, such as the effect of the conductor on emotional expression portrayed by the ensemble are considerably more difficult to study, and have scarcely been addressed by empiric studies. We chose, in the current book chapter, to focus exclusively on right hand conducting gestures, which are much more prototypically defined than left-hand movements, in order to enable a methodological evaluation of conducting gestures and their effect on the musical outcome. Conducting gestures introduce a transformation of sound to movement (on the conductor's side), and vice versa (at the ensemble's end of the information flow). Therefore, a detailed discussion on sound-to-movement cross-modal transformations will proceed, followed by a systematic discussion of the kinematics of conducting gestures, and their effect on the musical outcome.

10.2   Conducting and Cross-Modal Relationships

At the basis of conducting as a communicative process lies cross-modal "translation" involving different sensory modalities: the conductor's bodily motion, perceived visually by performers, affects the sound patterns performers produce—patterns themselves generated through the performers' bodily action. This communicative process may have various semiotic grounds. A conducting movement pattern may serve as a conventional symbol, whose denotation is based on arbitrary convention or agreement, rather than on any inherent relationship between symbol (the conductor's movement) and its interpretant (the performed musical output). However, while such conventional stipulations could perhaps have shaped some elementary conducting gestures, they cannot account for the wealth of non-conceptualized movements which convey subtle shadings of expression to performers—slight changes in tempo, minute yet systematic deviations from the metronomic beat, shadings of dynamic change, or gradations of articulation and accent—movements whose precise enactment shapes musical expression and distinguishes one performance from another.

One source of cross-modal "translation" in conducting may be the multi-modal nature of the temporal features communicated by a conductor. Features such as beat, meter or patterns of articulation may be depicted by both auditory and non-auditory stimuli (e.g., points of light, changes in tactile pressure, bodily motion), and correspondingly perceived through non-auditory sense modalities—visual, tactile, or proprioceptive (Guttman et al. 2005; Huang et al. 2012; Ross et al. 2016). Indeed, recent studies suggest that the same brain network was activated in beat perception tasks regardless of the sensory modality applied—auditory, visual, or tactile (Araneda et al. 2017). The cross-modal translation involved in conducting may employ such supra-modal resources.

Temporal features such as beat and meter are particularly associated with motor action, either overtly (tapping, dancing, walking, etc.) or covertly, by way of mental preparation and simulation of action. The latter is revealed by the activation of brain areas associated with preparation and support of motion, such as the dorsal premotor cortex and the supplementary motor area (SMA), during passive beat perception (see Ross et al. 2016, for research review). Notably, the association of auditory beat and meter with bodily motion is reciprocal. Just as musical beat and meter may induce corresponding bodily motion, movement expressing metrical structure may engender the perception of beat and meter in sound. For instance, 7-month old infants who were bounced every two or every three beats while listening to an ambiguous rhythm, later preferred a version of that rhythm corresponding to the metrical pattern (double or triple) induced by their bouncing (Phillips-Silver and Trainor 2005, 2007). The precise association of a conductor's movements with performed tempo and meter may rely on this strong, possibly innate ability to induce beat and meter through bodily motion.

Unlike such multimodal temporal features, dimensions such as loudness or pitch are essentially auditory. This notwithstanding, such auditory dimensions may exhibit consistently perceived correspondences with non-auditory features. Louder sound, for instance, is associated with larger objects, and changes in loudness—with motion in both the vertical and depth axes: crescendo with spatial rise and approach, and diminuendo—with fall and withdrawal. Such *cross-modal correspondences* (CMC)—"systematic associations found across seemingly unrelated features from different sensory modalities" (Parise 2016)—may affect basic aspects of human perception and cognition, including cross-modal binding in time and space (Parise and Spence 2009), perceptual learning (Brunel et al. 2015), selective attention (Marks 2004), and the perception of spatial location and motion direction (Pratt 1930; Maeda et al. 2004). Such effects are often automatic, and do not necessarily rely on conscious reflection or conceptual thought (for research reviews see Eitan 2013, 2017; Marks 2004; Spence 2011).

In the context of this paper, CMC of loudness or loudness change are of particular interest. Conductors convey loudness primarily through movement amplitude: a wider movement signifies a louder sound (see section on controlling loudness in conducting for more detailed discussion). Necessarily, increasing amplitude while maintaining a steady beat implies increasing velocity; hence, louder sound is also associated with faster movement. This practice reflects two well-established CMC involving loudness. First, greater loudness (or "volume") is associated with larger size. Children as young as 3 years old, as well as adults varying in cultural background, associate louder sound with larger physical size (Lipscomb and Kim 2004; Smith and Sera 1992; Walker 1987). This association may be based on the experienced correlation between the object size and the loudness of sounds it produces. Indeed, loudness significantly partakes in the auditory discrimination of objects' size (Burro and Grassi 2001). Second, loudness is associated with speed, as well as speed change. Louder music was rated as "faster" than softer music with the same tempo (Katz 2011). Loudness changes were associated with speed changes, both in music-induced imagery tasks (Eitan and Granot 2006; Eitan and Tubul, 2010) and in actual motion tasks, in which children reacted to crescendi and diminuendi with accelerating and decelerating body movements, respectively (Kohn and Eitan 2016).

Two additional CMC involving loudness partake in conductors' movements (expressed by left-hand gestures). First, loudness change is experientially (as well as acoustically) associated with change in distance. In particular, loudness increase (crescendo) serves as a fundamental signal for approach, affecting even infants rapidly and pre-attentively (see Eitan 2013, for a research review). Second, loudness change has been strongly associated with the vertical direction of motion. In fact, listeners' association of diminuendi (gradual loudness decrease) with spatial descent is as strong as their association of pitch descent and spatial descent (Eitan 2013). Even when not intentionally expressed by the conductor, these CMC may still affect performers' reactions.

Notably, a conductor does not merely indicate any *crescendo* or *diminuendo*, but rather, a particular change in loudness, with its specific amplitude envelope, duration, and extent. Quantitatively modeling how aspects of the conductor's motion suggested by loudness-related CMC, including amplitude, velocity, jerk or curvature, map onto aspects of performed loudness levels and loudness changes would be an interesting and necessary challenge for empirical studies of conducting.

Unlike loudness or tempo, musical dimensions such as pitch height or timbre are not explicitly conveyed by a conductor's movements. Nevertheless, one may inquire whether such dimensions may still affect these movements through implicit application of CMC. Consider pitch height, associated with several aspects of physical space and motion. Lower pitch, for instance, is associated with low and left-hand location, slow speed, and large physical size (Spence 2011; Eitan 2013). While Pitch height is normally not conveyed intentionally by a conductor's movements, CMC of pitch and space may still affect these movements. Moreover, CMC of pitch (and of other musical features not explicitly expressed by a conductor) may unintentionally affect aspects of a conductor's posture, head and neck movements, or facial expressions. Though not codified into an established repertoire of conducting movements, such effects may still influence performers' reactions. Exploring the effects of implicit CMC on a conductor's actions (including not only hand movements, but all visible body movements, postures, and facial expressions), and correspondingly investigating whether and how they are reflected in performers' output, could present an exciting new frontier for an empirical investigation of the art of conducting.

10.3 Beat

Right hand conducting gestures are primarily designed to communicate two main rhythmic parameters: beat and meter. The beat is the inner pulse of the music. It is a subjective, perceptual phenomenon, perceived as a regular pace determined by the pace of musical changes in time. The frequency of beats (i.e., the mean inter-beat-intervals per-unit of time) determines the perceived tempo of the music. The range of tempi (Italian: plural for tempo) usually involved in human music is between 40 and 300 Beats Per Minute (BPM), corresponding to inter-beat-intervals of 200–1500 ms (Van Noorden and Moelants 1999). The timing of musical beats is indicated in conducting gestures by reaching points in space which are characterized by distinct kinematic properties. Theoretically, one may expect the location of the beginning of musical beats in conducting gestures to correspond to the points of change in the direction of the movement trajectory. Surprisingly, this seemingly trivial property of conducting is controversial. A study by Clayton (Clayton 1986) demonstrated that the ictus (the point in time of the beginning of the beat) is determined by the lowest part of the trajectory along the vertical axis, either as the baton rounds the corner, or bounces back up. Luck (2000) corroborated these findings, but also hypothesized that the perception of timing of the ictus is affected by multiple kinematic factors (G. Luck 2000). Later studies supported this supposition, showing that synchronicity in the orchestra's playing is strongly associated with the acceleration of the baton hand (Luck 2008; Luck and Sloboda 2009; Luck and Toiviainen 2006). Accordingly, the perceived location of a beat was believed to be based on the position landmark of the maxima in vertical acceleration. This finding was also validated by Wöllner et al. (2012), who investigated the reaction of musicians to morphed conducting gestures. Their findings indicated that musicians find it easier to synchronize with morphed gestures (the grand averaged conducting gestures of 12 different conductors) than with real-life, individual conducting gestures. A preference for morphed gestures may indicate that orchestra musicians form an internal model of average, prototypical set of conducting gestures, enabling them to respond quickly to the great variety of conductors they encounter as professional orchestral players, as long as their conducting gestures are in line with the "prototypical" conducting technique (Wöllner et al. 2012). Additional findings indicate that synchronization among musicians is enhanced when conducting gestures resemble a kind of motion, similar to that produced by gravity (Luck and Sloboda 2007). These findings are actually not in line with the ictus being at the point of maximum acceleration, since free fall movements involve constant acceleration. Since the direction of movement is changed at the end of the downbeat, it is actually the point of maximum deceleration which might be easier to detect. These findings are in line with the results obtained by Luck and Toiviainen (2006), who showed that maximal synchronicity within the ensemble was reached in points of maximal deceleration. They also pointed at vertical velocity as an important factor influencing ensemble synchronicity (Luck and Toiviainen 2006). Luck and Sloboda (2007) pointed out that gravity causes higher downward velocity, necessitating a more marked deceleration as the gesture rounds the corner (Luck and Sloboda 2007). These results indicate enhanced changes in velocity towards the 1st beat compared to all other beats, predicting more pronounced ensemble synchronization for the downbeat (see next section for a detailed discussion on synchronization with the meter). Another relevant finding shows a linkage between gesture curvature and synchronicity in the ensemble's response. Luck (2008) showed that synchronicity is negatively correlated with the amount of curvature contained within a gesture (Luck 2008). This adds curvature to the factors influencing conductor-ensemble synchronization and interpretation of the ictus.

The discrepancies between findings regarding the kinematic characterization of the ictus could derive from several factors (or their combination): a. Diversity of conducting techniques between conductors who participated in different studies. b. Differences in experimental methodologies and in analysis methods. c. Differences in synchronization to different beats in the meter, d. Differences in experience of following various conductors who took part in a large array of different studies. Supportive of the last factor are results showing that synchronicity within the orchestra is positively correlated with experience, introducing an additional factor into the equation. Another finding by Luck and Toiviainen (2006) showed that musicians respond with a significant delay between the perceived ictus and the actual onset of the ensemble (Luck and Toiviainen 2006). Interestingly, musicians tend to lag behind the beat more than non-musicians. These findings highlight the difficulty encountered in attempting to track down the moment of the ictus. A variable delay in response to the ictus requires a recalculation of the results, based on orchestral experience. Notwithstanding the methodological difficulties in characterizing the point of ictus, there seems to be an agreement on the importance of acceleration (and deceleration) as a major factor influencing ictus detection.

10.4 Meter

Meter is defined as the pattern in which a steady succession of rhythmic pulses are organized (Harvard Dictionary of Music: (The new Harvard dictionary of music 2003). The organization of the beats in a meter is hierarchically defined, with the first beat (the "downbeat") in the meter being the most accentuated. This principle is clearly reflected in dance music, where the meter reflects the patterns of dancing steps, and their periodic nature. For example, the dance pattern of a waltz includes a periodic pattern of three steps, which is also conveyed by the music. Sub-hierarchies also exist within a meter containing more than 3 inner beats. For example, in a meter of four, the first beat is the most accentuated, the third beat contains a lighter accent than the first beat, while the second and fourth beats are perceived as lighter. This hierarchical structure divides the meter of four into three sub-hierarchies. Similarly, a meter of 5 can be hierarchically divided into two successive groups of $2 + 3$ or $3 + 2$, a meter of 6 into two groups of $3 + 3$, etc. Conducting gestures are specifically designed to communicate these hierarchies, through division in the space of these hierarchical groups. The "heaviest" beat in a meter would always be the first one (the "downbeat") which is conducted with a downward movement, possibly signifying the heaviness of the force of gravity. The last beat of a meter (the "upbeat"), accumulates musical energy towards the first, heavy beat, and is conducted with an upward movement, possibly symbolizing an increase in potential energy and tension. The rest of the beats in any meter, are indicated by a movement to the right or the left (see Fig. 10.1 for a detailed description of right-hand conducting

Beat

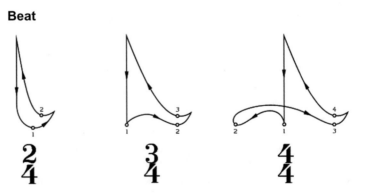

Fig. 10.1 Conducting gestures for 2, 3, and 4 beat meters

gestures), parallel to the ground. There is no formal explanation for the choice of direction of the beats. However, one could speculate that there might be a kinematic-musical explanation for choosing specific directions for specific beats in a meter. For example, in a meter of 4, the third beat, which is partially accentuated, is conducted in an outward movement, which allows a greater movement-amplitude, signaling louder sound production. As discussed in the previous section, conductor-ensemble synchronization tends to be at its highest level on the downbeat of the meter (Luck and Sloboda 2007). There may be several explanations for this phenomenon. The most obvious is the change in movement direction characterizing both the upbeat and downbeat in conducting gestures. An additional explanation derives from multiple studies demonstrating periodic allocation of attention, affected by metric hierarchies. Attention level appears to be at peak towards the downbeat (Fitzroy and Sanders 2015). Taken together, it is possible that the greater synchronicity of ensembles on the downbeat of the meter derives both kinematic factors and periodic allocation of attention. This conjecture requires further examination in empiric studies, evaluating the effect of meter on visual perception of movement.

10.5   Loudness

Loudness is one of the most important cues indicating character in music (Juslin 2000; Juslin and Laukka 2004). Classical music compositions include detailed instructions regarding the degree of loudness (e.g. forte and piano-Italian terms for loud and soft). Loudness is conveyed in conducting through the amplitude of the movement, both in theory and in practice. The larger the movement, the louder the musical outcome (Wöllner et al. 2012). Very often in music, a steady beat is maintained, while loudness changes. In this case, spatial inter-beat intervals of conducting gestures remain constant, while amplitude and velocity will change to convey the change in loudness. This alteration of movement kinematics is carried out effortless by novices in conducting. A simple exercise recommended for the reader, demonstrates the easiness with which such alteration of movement dynamics is carried out. Tapping at a steady beat, with increasing amplitude of tapping movement (which will result in louder taps) should prove to be of no difficulty for anyone in the general population with no musical background. The "principle of isochrony" could prove a credible explanation for this phenomenon. This principle indicates that the duration of movements remains nearly constant, while changing amplitude, speed and acceleration. In other words, movement duration is nearly independent of its amplitude. Two aspects of isochrony were reported: 'global isochrony' (Viviani and Schneider 1991) refers to the near constancy of the entire movement duration, while 'local isochrony', refers to the observation that within more complicated movements the durations of the internal segments are also roughly equal independently of the segments' lengths, as long as the entire movement is planned as a whole (Flash and Hogan 1985). Going back to conducting, the principle of isochrony enables conductors to easily convey expression through change in amplitude, muscle tension, curvature of movement etc.

while maintaining a steady beat. The same would probably apply to the ensemble, having no difficulty in keeping the pulse, while extracting relevant additional information on loudness, sound quality, articulation etc. conveyed by various kinematic properties of conducting gestures, which are independent of movement duration.

10.6 Articulation

Articulation can be defined as "The characteristics of attack and decay of single tones or groups of tones, and the means by which these characteristics are produced" (The new Harvard dictionary of music 2003). Another definition by the Grove Dictionary reads: "The manner in which successive notes are joined to one another by a performer" (Grove Dictionary of Music and Musicians 1995). A unified definition for articulation would encompass both duration and attack. Articulation has a substantial effect on the emotional impact of musical performance (Juslin 2000). For example, while sadness is associated with legato (tied) articulation, happiness is associated with staccato (separated and short) articulation. Hence, communicating a complete emotional characterization of the music requires clear instruction regarding articulation to be included in conducting gestures. The acoustic manifestation of sound attack and decay can be observed in 4 main stages of loudness progression of sound in time: attack, decay, sustain and release (Hähnel and Berndt 2010). An accentuated attack would be reflected in an abrupt decay at the attack stage of the sound, while a softer attack would be reflected in a slower decay. Conducting gestures can portray these acoustic changes through jerk (the rate of change of acceleration). The more jerky a movement is, the more accentuated (i.e.-played with an accent) the sound (Nakra et al. 2009).

The duration of sound is rather simple to control through conducting gestures. Stopping the movement of the right-hand stops sound. This straightforward technique is described in the following quotation from McElheran's method of conducting: "*Simply hold the hand still when you come to the fermata, letting the note continue as long as you wish*" (McElheran 2004). Short tones ("staccato") are represented in conducting technique by a quick wrist movement, signifying the quick bounce of the bow, in string players, or the tongue in wind players (Rudolf 1980, p. 16).

10.7 Motion, Emotion and Sound: Expressive Conducting, Unraveling the Enigma

One of the most enigmatic features of conducting is the ability of conductors to inspire the ensemble, at times leading to an enrapturing performance. This intriguing feature of conducting is difficult, if not impossible to quantify. Conducting gestures alone cannot possibly account for the great differences in musical character obtained by different conductors. There is, no doubt, a central role for personal charisma, facial expression, professional authority, quality of verbal instruction, as well as many other personal and professional traits underlying the ability to successfully transmit an emotional message to an ensemble. However, it is possible that one could isolate some of the kinematic characteristics involved in communicating emotional messages in conducting. Prior studies investigating the expressive content of human motion may provide important information regarding conductor-ensemble emotional communication. Pollick et al. (2001) investigated point-light displays of human arm movements and their ability to convey an emotional message to spectators (Pollick et al. 2001).They employed the circumplex model of emotion to evaluate the emotional impact of movements (Russell 1980). This model (as well as other multidimensional models of emotion) suggests that emotions can be described by a two-dimensional circular space: activation, represented by the vertical axis, and valence, represented by the horizontal axis. Their results indicated that the amount of activation characterizing the emotional content of arm movements is positively correlated with velocity, acceleration and jerk of the movement.

As mentioned before in this chapter, movement's jerk is also an efficient method of conveying articulation (Nakra et al. 2009). Interestingly, the motor system does not show an inclination towards larger jerk. Rather the contrary: mathematical models based on optimization theory (Flash and Hogan 1985; Harris and Wolpert 1998; Todorov and Jordan 1998, 2002; Uno et al. 1989) suggest that the motor system optimizes certain costs, those being based on either kinematic or dynamic variables or variables related to neural activations. According to the minimum-jerk model (Flash and Hogan 1985), a primary objective of motor coordination is to achieve the smoothest possible hand trajectory under the circumstances. This objective was equated with the minimization of hand jerk (the rate of change of hand acceleration), and the specific trajectories yielding the optimal performance were mathematically determined and compared to measured movements. Hence, jerky conducting movements are less consistent with the optimal characteristics of movement. Furthermore, one would expect experienced conductors to reconcile between the principle of jerk minimization (i.e., reducing movement jerk to a minimum), and the necessity to convey articulation and emotional activation, reaching a result which would seem more natural to the musicians and audience alike. This conjecture calls for further investigation, focusing on jerkiness and gesture effectiveness in novice and experienced conductors.

Timing and amplitude were also found to correlate with activation (Amaya et al. 1996). While arm movement kinematics seems to portray activation reliably, valence does not seem to be portrayed as clearly. It is now assumed that valence is conveyed through a multimodal combination of body postures, facial expression and movement kinematics. Darwin, In "The Expression of Emotion in Man and Animals", (1872) emphasized the importance of body postures in conveying emotions both in animals and in humans. These views have been supported by a multitude of studies, demonstrating the emotional impact of body posture on human spectators. It is beyond the scope of the current book chapter to discuss static body postures, and their combination with movement kinematics in conducting. However, it seems imperative for future studies, investigating the emotional impact of conducting on the ensemble and audiences, to include both static and dynamic features in their analysis of conducting.

10.8 Conclusion

Despite the growing number of studies on conducting gestures and their effect on musical performance, conducting remains an enigmatic phenomenon. We strongly believe that the fascinating interaction between conductor and ensemble is a holistic phenomenon, which can only be understood by combining multidisciplinary research methods, including social psychology, computational motor control (in action and perception, alike), musicological analysis tools, physics and acoustics of musical instruments, biomechanics of conducting and music-instrument playing, as well as cross-modal sound-to-movement mapping.

References

Amaya, K., Bruderlin, A., & Calvert, T. (1996). *Emotion from motion.* Toronto, Ontario: Paper presented at the Graphics Interface.

Araneda, R., Renier, L., Ebner-Karestinos, D., Dricot, L., & De Volder, A. G. (2017) Hearing, feeling or seeing a beat recruits a supramodal network in the auditory dorsal stream. European Journal of Neuroscience, 45(11), 1439–1450.

Brunel, L., Carvalho, P. F., & Goldstone, R. L. (2015). It does belong together: Cross-modal correspondences influence cross-modal integration during perceptual learning. *Frontiers in Psychology, 6,* 358. https://doi.org/10.3389/fpsyg.2015.00358.

Burro, R., & Grassi, M. (2001). Experiments on size and height of falling objects. In *Phenomenology of sound events*, IST project no. IST-2000–25287, report no. 1 (pp. 31–39).

Clayton, A. M. H. (1986). Coordination Between Players in Musical Performance. (Ph.D.), Edinburgh University, Edinburgh.

Eitan, Z. (2013). How pitch and loudness shape musical space and motion: New findings and persisting questions. In S. L. Tan, A. Cohen, S. Lipscomb, & R. Kendall (Eds.), *The psychology of music in multimedia* (pp. 161–187). Oxford: Oxford University Press.

Eitan, Z. (2017). Cross-modal correspondences. In Ashley, R., & Timmers, R. (Eds.), *The routledge companion to music cognition* (pp. 213–224). Routledge.

Eitan, Z., & Granot, R. Y. (2006). How music moves: musical parameters and images of motion. *Music Perception, 23,* 221–247.

Eitan, Z., & Tubul, N. (2010). Musical parameters and children's images of motion. *Musicae Scientiae, Special Issue,* 89–111.

Fitzroy, A. B., & Sanders, L. D. (2015). Musical meter modulates the allocation of attention across time. *Journal of Cognitive Neuroscience, 27*(12), 2339–2351. https://doi.org/10.1162/jocn_a_00862.

Flash, T., & Hogan, N. (1985). The coordination of arm movements: An experimentally confirmed mathematical model. *Journal of Neuroscience, 5*(7), 1688–1703.

Grove Dictionary of Music and Musicians. (1995). Oxford: Oxford University Press.

Guttman, S. E., Gilroy, L. A., & Blake, R. (2005). Hearing what the eyes see: Auditory encoding of visual temporal sequences. *Psychological Science, 16*(3), 228–235.

Hähnel, T., & Berndt, A. (2010). *Expressive articulation for synthetic music performances.* Paper presented at the Proceedings of New Interfaces for Musical Expression, Sydney, Australia.

Harris, C. M., & Wolpert, D. M. (1998). Signal-dependent noise determines motor planning. *Nature, 394*(6695), 780–784.

Huang, J., Gamble, D., Sarnlertsophon, K., Wang, X., & Hsiao, S. (2012). Feeling music: Integration of auditory and tactile inputs in musical meter perception. *PLoS ONE, 7*(10), e48496.

Juslin, P. N. (2000). Cue utilization in communication of emotion in music performance: Relating performance to perception. *Journal of Experimental Psychology: Human Perception and Performance, 26*(6), 1797–1813. https://doi.org/10.1037//0096-1523.26.6.1797.

Juslin, P. N., & Laukka, P. (2004). Expression, perception, and induction of musical emotions: A review and a questionnaire study of everyday listening. *Journal of New Music Research, 33*(3), 217–238. https://doi.org/10.1080/0929821042000317813.

Katz, A. (2011). *Metaphor as representation of children's musical thought: Metaphorical mapping and musical parameters.* Doctoral dissertation, Tel Aviv University. (In Hebrew).

Kohn, D., & Eitan, Z. (2016). Moving music: Correspondences of musical parameters and movement dimensions in children's motion and verbal responses. *Music Perception, 34*(1), 40–55.

Lipscomb, S. D., & Kim, M. (2004). *Perceived match between visual parameters and auditory correlates: An experimental multimedia investigation.* Paper presented at the 8th International Conference on Music Perception and Cognition. Northwestern University in Evanston, Illinois, USA, August 3–7, 2004.

Luck, G. (2000). *Synchronizing a motor response with a visual event: The perception of temporal information in a conductor's gestures.* Paper presented at the Sixth International Conference on Music Perception and Cognition, Keele University, UK.

Luck, G. (2008). Conductors' temporal gestures: Spatiotemporal cues for visually mediated synchronization. *Journal of the Acoustical Society of America, 124*(4), 2447.

Luck, G., & Sloboda, J. A. (2007). An investigation of musicians' synchronization with traditional conducting beat patterns. *Music Performance Research, 1,* 26–46.

Luck, G., & Sloboda, J. A. (2009). Spatio-temporal cues for visually mediated synchronization. *Music Perception, 26*(5), 465–473.

Luck, G., & Toiviainen, P. (2006). Ensemble musicians synchronization with conductors gestures: An automated feature-extraction analysis. *Music Perception, 24*(2), 189–200.

Maeda, F., Kanai, R., & Shimojo, S. (2004). Changing pitch induced visual motion illusion. *Current Biology, 14,* R990–R991.

Marks, L. (2004). Cross-modal interactions in speeded classification. In G. Calvert, C. Spence, & B. Stein (Eds.), *The handbook of multisensory processes* (pp. 85–106). Cambridge, MA: MIT Press.

McElheran, B. (2004). *Conducting technique: For beginners and professionals.* USA: Oxford University Press.

Nakra, T. M., Salgian, A., & Pfirrmann, M. (2009). *Musical analysis of conducting gestures using methods from computer vision.* Paper presented at the international computer music conference, Montreal, Canada.

Parise, C. V. (2016). Crossmodal correspondences: Standing issues and experimental guidelines. *Multisensory Research, 29*(1–3), 7–28.

Parise, C. V., & Spence, C. (2009). 'When birds of a feather flock together': Synesthetic correspondences modulate audiovisual integration in non-synesthetes. *PLoS ONE, 4,* e5664. https://doi.org/10.1371/journal.pone.0005664.

Phillips-Silver, J., & Trainor, L. J. (2005). Feeling the beat: Movement influences infant rhythm perception. *Science, 308*(5727), 1430.

Phillips-Silver, J., & Trainor, L. J. (2007). Hearing what the body feels: Auditory encoding of rhythmic movement. *Cognition, 105*(3), 533–546.

Pollick, F. E., Paterson, H. M., Bruderlin, A., & Sanford, A. J. (2001). Perceiving affect from arm movement. *Cognition, 82*(2), B51–B61.

Pratt, C. C. (1930). The spatial character of high and low tones. *Journal of Experimental Psychology, 13,* 278–285.

Ross, J. M., Iversen, J. R., & Balasubramaniam, R. (2016). Motor simulation theories of musical beat perception. *Neurocase, 22*(6), 558–565.

Rudolf, M. (1980). *The grammar of conducting: A practical study of modern baton technique.* New York: Schirmer Books.

Russell, J. (1980). A circumplex model of affect. *Journal of Personality and Social Psychology, 39*(6), 1161–1178. https://doi.org/10.1037/h0077714.

Scherchen, H. (1989). *Handbook of conducting.* New York Oxford University Press.

Schuller, G. (1999). *The compleat conductor.* New York: Oxford University Press.

Seidl, A. (1899). On conducting. In H. T. Finck (Ed.), *Anton Seidl: A memorial by his friends.* New York: Charles Scribners Sons.

Smith, L. B., & Sera, M. D. (1992). A developmental analysis of the polar structure of dimensions. *Cognitive Psychology, 24,* 99–142.

Spence, C. (2011). Crossmodal correspondences: A tutorial review. *Attention, Perception, & Psychophysics, 73*(4), 971–995.

The new Harvard dictionary of music. (2003). *Cambridge.* Massachusetts: Harvard University Press Reference Library.

Todorov, E., & Jordan, M. I. (1998). Smoothness maximization along a predefined path accurately predicts the speed profiles of complex arm movements. *Journal of Neurophysiology, 80*(2), 696–714.

Todorov, E., & Jordan, M. I. (2002). Optimal feedback control as a theory of motor coordination. *Nature Neuroscience, 5*(11), 1226–1235.

Uno, Y., Kawato, M., & Suzuki, R. (1989). Formation and control of optimal trajectory in human multijoint arm movement. *Biological Cybernetics, 61*(2), 89–101.

Van Noorden, L., & Moelants, D. (1999). Resonance in the perception of musical pulse. *Journal of New Music Research, 28*(1), 43–66.

Viviani, P., & Schneider, R. (1991). A developmental study of the relationship between geometry and kinematics in drawing movements. *Journal of Experimental Psychology: Human Perception and Performance, 17*(1), 198–218.

Walker, R. (1987). The effects of culture, environment, age, and musical training on choices of visual metaphors for sound. *Perception and Psychophysics, 42,* 491–502.

Wöllner, C., Deconinck, F. J., Parkinson, J., Hove, M. J., & Keller, P. E. (2012). The perception of prototypical motion: Synchronization is enhanced with quantitatively morphed gestures of musical conductors. *Journal of Experimental Psychology: Human Perception and Performance, 38*(6), 1390.

Chapter 11
Interaction, Cooperation and Entrainment in Music: Experience and Perspectives

Luciano Fadiga, Serâ Tokay, and Alessandro D'Ausilio

Abstract Complex multi-agent behavioral coordination requires the capability of reading and sending subtle sensorimotor messages while performing a joint action towards a shared goal. While progress has been made in the field of social neuroscience, the neurobehavioral mechanisms underlying this kind of interaction, still represent the 'dark matter' of cognitive neuroscience. Here we present a series of investigations using ensemble musicians as a test-bed to explore whether motion kinematics and advanced time-series analysis could be used to extract these dynamics. The data we report suggests that the pattern of sensorimotor communication flow between musicians and conductors modulate joint action outcome. Furthermore, we also demonstrate that music ensemble communication is conveyed by movements of different body parts, each one of them containing complementary information needed for coordination. This research line has thus the potential to unravel the multi-scale and multi-channel nature of human sensorimotor communication.

11.1 Introduction

All animal species grouping for defensive, reproductive or hunting needs have evolved complex communicative behaviors to achieve coordinated action (Frith 2008; Rands et al. 2003; Couzin et al. 2005; Nagy et al. 2010). Humans are innately social creatures and there is little doubt that cognition and brain organization are shaped around this fact. Nevertheless, cognitive neuroscience took quite some time to acknowledge that the study of cognition in isolated individuals might be an ill-posed scientific approach. Recently however, a clear switch towards what has been called "second person neuroscience" has led to an apparent blossoming of the study

L. Fadiga (✉) · S. Tokay · A. D'Ausilio
Center for Translational Neurophysiology, Istituto Italiano di Tecnologia, Ferrara, Italy
e-mail: fdl@unife.it

Section of Physiology, University of Ferrara, Ferrara, Italy

© Springer Nature Switzerland AG 2021
T. Flash and A. Berthoz (eds.), *Space-Time Geometries for Motion and Perception in the Brain and the Arts*, Lecture Notes in Morphogenesis,
https://doi.org/10.1007/978-3-030-57227-3_11

of the brain during real-time truly interactive scenarios (Schilbach et al. 2013; Hari et al. 2015).

A major theoretical shift in this regard was driven by a new conceptualization of the motor system. The motor system was once believed to be an output system, slavishly following the dictate of the perceptual brain. However, motor processes seem to also play a role in perceptual and cognitive functions, challenging the classical sensory versus motor separation and opening the doors to embodied cognition research in both humans (Clark and Grush 1999) and artificial systems (Vernon et al. 2011). In addition to the fundamental role in movement planning and execution, some premotor neurons show also complex visual responses. Among them, "mirror neurons" discharge both when the monkey executes an action and observes another individual make the same action (Gallese et al. 1996). On the other hand, "canonical neurons" are activated both during object-directed actions and object observation alone (Murata et al. 1997). These discoveries have stimulated a large wealth of basic monkey and human research and after about 20 years we know a great deal of details regarding the role and function of the motor system in action and object perception (D'Ausilio et al. 2015a).

In fact, action properties and object geometry carry reliable statistical features we can exploit for classification. Indeed, actions must comply with the laws of biological motion and body biomechanics, whereas the shape of objects/tools constrain the interaction potentialities allowed to the human hands. All this a priori knowledge offers a great advantage for any system that tries to optimize behavioral interaction. In this sense, activities in the motor system are not intended as a coding schema of the external world but rather as a generative engine to extract different levels of features depending on context (D'Ausilio et al. 2015b). Critically, the recruitment of motor programs, during action/object perception, constrain the active search of specific sensory features that maximize the discrimination between two perceptual hypotheses and/or support prediction of future information (Friston et al. 2011).

The generation of active inferences about future actions of conspecifics is central to our capability to smoothly interact with each other and, therefore, fundamental to the development of human cognition. An interesting research theme revolves around the investigation of how these mechanisms are instantiated at the brain level and how these processes are used to facilitate human-to-human interaction. In simple terms, the quest for the understanding of how, why and when we do send and receive implicit sensorimotor messages during complex social interaction.

11.2   The Neuro-functional Substrate of Sensorimotor Communication

This sensory-motor conversation requires that all participants must be able to send and receive subtle messages in the form of visual motor gestures and auditory events. Therefore, critical for this account is the ability to encode/decode such messages. This encoding/decoding process might be conceived as a complex and hierarchical input-output mapping ranging from rote sensory-motor mapping to the highest level of human action organization (Grafton and Hamilton 2007; Wolpert et al. 2003). The fronto-parietal networks are the best neural candidates for this computation since they are characterized by extensive structural connections (Petrides and Pandya 2009) supporting critical higher-level sensori-motor coordinate transformations (Rizzolatti and Luppino 2001). The circuit connecting the anterior intraparietal sulcus (AIP) and the posterior bank of monkey inferior arcuate sulcus (F5), for instance, has shown interesting functional properties (Borra et al. 2008; Rozzi et al. 2006; Luppino et al. 1999). A subset of neurons in both these areas was activated by the execution of goal-directed actions and the observation of kinematically similar actions performed by other individuals (Gallese et al. 1996; Fogassi et al. 2005).

Originally found in monkeys by using single unit recordings, subsequent studies have shown similar mechanisms in humans as well. Different techniques such as transcranial magnetic stimulation (Fadiga et al. 1995), fMRI or PET (Rizzolatti et al. 1996; Decety et al. 1997; Buccino et al. 2001; Grezes and Decety 2001) and MEG (Hari et al. 1998) have confirmed these results. More recently, a large body of research is also describing these mechanisms in humans by using novel methods and approaches (Avenanti et al. 2007; Gazzola and Keysers 2009; Fazio et al. 2009; Kilner et al. 2009; D'Ausilio et al. 2015a).

The mirror fronto-parietal circuit, due to its properties, might be particularly important in communication between individuals, especially when coordinated action is central (Hurley 2008; Sommerville and Decety 2006; Rizzolatti and Craighero 2004). In fact, the other's action might be first represented in our motor system via mirror-like mechanisms, such that we can easily have access to "how" the agent is doing what he's doing (Fadiga et al. 1995; Borroni et al. 2005). This knowledge, on one hand, will help us understand the agent's intentions via additional inferential processes eventually based on experience and contextual information (Iacoboni et al. 2005). On the other hand, instead, this low level sensory-motor resonance with our own motor planning repertoire might enable efficient prediction, anticipation and planning of appropriate actions in response or modify our behavior online (Iacoboni et al. 1999).

Action mirroring, however, does not facilitate coordinated action per se, in fact, coordination may often require the execution of very different actions between participants. For example, if we move a heavy load together with somebody else we might be better tracking the other's action in a fast and efficient manner. In fact, if our fellow produces a particular movement in one direction we need to compensate with a concurrent and opposite action that takes into account several kinematic and dynamic aspects of the situation. This is a clear example of coordinated action where one individual is temporarily causing the others' behavior (Sebanz et al. 2006). Recently however, it has been shown that complementary action observation may recruit human mirror neuron areas to a greater extent (Newman-Norlund et al. 2007) thus suggesting that the human mirror system might be tuned to action coordination rather than rote action mirroring. We propose that the mirror-mechanism might provide the means to represent the other's action in motor coordinates rather than

visual or symbolic (i.e. language) that in turn may result extremely useful in anticipating other's actions, planning a motor correction and hence be useful for fast
coordinated behaviors. In this context, we might use our motor internal models to
gain access to low-level control parameters implemented by the people interacting
with us. Therefore, action coordination may benefit from the ability to model others'
behavior implementation and use it as an additional prior, in a Bayesian perspective
(Friston et al. 2011).

11.3   The Use Musicians to Study Sensorimotor Processes

While engaged in naturalistic joint action, action coordination may be subject to data
availability (full body vision Vs occluded vision), context richness (known vs novel
context) or expertise in a given task (Rizzolatti and Sinigaglia 2010). If on one hand
it is clear that partly occluded vision or the presence of a reliable context reduces
the impact of low level decoding of others' action kinematics, the case of experts is
particularly interesting. In fact, sport, dance or music experts do not obtain a simple
rough interpretation of others' actions. Rather, expertise in these activities is all about
anticipatorily modeling of low-level movement features (Aglioti et al. 2008).

   In fact, professional musicians, for instance, undergo important plastic changes
induced by extensive motor and sensory training (Elbert et al. 1995; Schlaug et al.
1995; Pantev et al. 1998; D'Ausilio et al. 2006). These anatomo-functional changes
are also paralleled by enhanced ability to discriminate subtle changes in others'
performance via predictive action simulation (Knoblich and Flach 2001; Wilson and
Knoblich 2005; D'Ausilio et al. 2010; Candidi et al. 2014). One facet of real expertise may be the refinement of sensori-motor skills as well as the domain-specific
tuning of low-level sensori-motor feature extraction from others behavior. In this
vein, musicians have been extensively used with these purposes. For example, drum-
mers were used to study time processing (Cicchini et al. 2012), violinist players for
somatosensory plasticity (Elbert et al. 1995), piano players for sensori-motor integration (D'Ausilio et al. 2006) or jazz players to study creativity and improvisation
(Limb and Braun 2008).

   For instance, listening to music evokes specific sensorimotor activity in musicians.
This fact has been shown using various neuroimaging and neurophysiological techniques (MEG: Haueisen and Knösche 2001; EEG: Bangert and Altenmüller 2003;
fMRI: Lotze et al. 2003; Bangert et al. 2006; Lahav et al. 2007; TMS: D'Ausilio
et al. 2006; Novembre et al. 2012). These results broadly suggest that a fronto-
parietal network of brain areas, usually activated by action planning and execution,
is, at least partially, recruited in the musicians' brain when listening to rehearsed
musical segments. Similarly, the visual presentation of musical actions (e.g. videos
of hand playing a piano) evokes specific motor brain responses (Haslinger et al. 2005;
Engel et al. 2012; Candidi et al. 2014), suggesting that musical training also generates new visuo-motor associations. Importantly, these sensorimotor activities are
triggered by sensory stimuli, even when no motor response is needed (Zatorre et al.

2007), suggesting that long-term training produces stable sensorimotor associations. Therefore, the motor system in musicians show mirror-like properties and it seems plausible that this mechanism could support their orchestrated musical execution.

11.4 The Use of Ensemble Musicians to Study Sensorimotor Communication

Recently it has been proposed that music can offer a unique solution to balance experimental control and ecological testing of cognition and brain functions (D'Ausilio et al. 2015a). More interestingly, musicians especially ensemble instrumentalists, are experts in a form of social interaction characterized by real-time non-verbal communication. Ensemble musicians train for years in order to refine skills that allow them to accurately encode and decode subtle sensorimotor non-verbal messages with the main purpose of establishing and maintaining a shared coordinative goal. In group level musical coordination individuals might be conceptualized as processing units embedded within a complex system (i.e. the ensemble) and sharing an aesthetic and emotional goal. Each participant may thus non-verbally transmit sensory information while in parallel decoding other's movement. The sensory information generated by the sender is based on body movements and is transferred through the visual (i.e., body sway, had motion), auditory (i.e., instrument sounds) and somatosensory channels (i.e., floor vibrations). With information flowing, participants may rely on predictive models to cope with the real-time demands of interpersonal coordination. Thus, the musical ensemble constitutes a dynamical system that possesses important constraints due to the fact that co-performers behave in a complex but formalized manner dictated by musical conventions and often a musical score.

As a matter of facts, for musical ensemble playing the complexity of behavioural coordination can be formalized as the accurate control of relative movement timing during performance. For example, quartets exhibit a near-optimal gain, whereas individual gains reflected contrasting strategies of first-violin-led autocracy versus democracy group organization (Wing et al. 2014). Furthermore, the familiarity with a co-performer's part has dissociated effects on different levels of movement control. Keystroke coordination was affected by predictions about expressive micro timing, based on the performer's own playing style rather than the co-performer's style. Body sway coordination was, instead, modulated by temporal predictions related to musical phrase structure (Ragert et al. 2013). These studies are starting to shed some initial light on the possibility that complex behavioral coordination could be based on a hierarchy of body motion control, each differentially modulated by social and/or situational factors. In this sense, ensemble music seems to be an effective test-bed for such a domain-general capability.

On the other hand, a series of studies based on motion kinematic data acquired from quartets or orchestras, used a complementary computational approach. These studies, recorded musicians playing in an ecological scenario and employed mathematical tools to extract natural information flow between participants. These data-driven techniques, such as the Granger Causality method, showed that the increase of conductor-to-musicians influence, together with the reduction of musician-to-musician coordination was associated with quality of execution, as assessed by musical experts (D'Ausilio et al. 2012; Fig. 11.1). A similar methodology was also able to extract implicit dominance and leadership in quartets (Badino et al. 2014), whereas entropy based measures distinguished between solo and ensemble performances (Glowinski et al. 2013). Together, these studies suggest that using specific computational tools it is possible to blindly extract the pattern of information transfer between participants engaged in a realistic and complex interaction—without any variables manipulation and behavioural interference.

According to the mirror-matching hypothesis, one possible mechanism supporting behavioural coordination via sensorimotor information transfer is the motor simulation of other's behaviour. For instance, the role of motor simulation in inter-personal coordination was investigated in pianists listening to a recording of a piece trained with the left hand, and asked to play a complementary part with the right hand. Interference with the motor cortex, impaired inter-personal coordination only when the heard pieces had been trained (Novembre et al. 2012). Of further interest is the finding that the mere presence of a co-performer modulated cortico-spinal excitability in amateur pianists asked to perform the right-hand part of a musical piece, while the complementary left-hand part was believed to be performed by a hidden co-performer (Novembre et al. 2012). This kind of studies might constitute a powerful tool to study the neurophysiological role of other's action simulation during dynamic interaction and behavioural coordination.

At the same time other recent studies started the investigation of musical interaction by complementing behavioural measures with classical metabolic and electrical neuroimaging. For instance, it was shown that simultaneous electroencephalographic recording in musicians playing in ensemble could be achieved with a limited amount of motion artefacts (Babiloni et al. 2012). Following this approach it was possible to dissociate the processes related to monitoring the self's performance and the joint action outcome (Loehr et al. 2013). Even more compelling is the possibility to investigate the amount of inter-brain information flow. In fact, patterns of directed between-brain coupling, in alpha and beta frequency ranges, was associated with the musical roles of leader and follower (Sänger et al. 2013). Also, the increase of inter-brain phase coherence in delta and theta frequency bands was increased in frontal and central scalp regions during periods that presented higher demands on coordination (Sänger et al. 2012).

Functional Magnetic Resonance Imaging showed the neural correlates of inter-personal synchrony and its effect on pro-social behaviour. Synchronization was shown to increase activity in the caudate (reward system) and predicted prosodical behaviour (Kokal et al. 2011). Another study identified distinct cortical networks that were selectively activated depending on the kind of cooperativeness of an interaction partner. Cooperative virtual partners who corrected for moderate amounts of

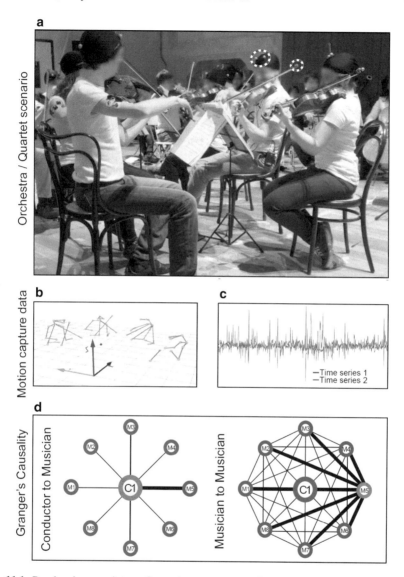

Fig. 11.1 Panel **a** shows a picture of a motion capture recording set-up. Highlighted the markers on the musicians head and bows. In panel **b**, a stick-figure representation of a quartet performin a musical piece. In panel **c**, two time-series representing accelerometric data from a motion capture recording. In panel **c**, a graphical description of the metrics that can be extracted from movements. Specifically, on the left side what we called Conductor to Musicians" describe the causal drive from the conductors towards all musicians. On the right side, what we named "Musicians to Musicians", an index that while removing the statistical contribution of the conductor measured the inter-musician sensorimotor communication

synchronization error facilitated performance and were associated with the activation of cortical midline structures linked to socio-affective processes. Virtual partners who over-compensated for errors led to poor inter-agent synchronization and the activation of lateral prefrontal areas associated with executive functions and cognitive control (Fairhurst et al. 2013). Also, leader-follower dynamics in inter-agent coordination showed that agency related brain regions were strongly activated in leaders, consistent with their greater self-focus (Fairhurst et al. 2014).

In summary, the basic building blocks enabling social dominance, leadership, or cooperativeness might be effectively be studied in ensemble musicians through the use of multi-dimensional approaches (involving combinations of brain imaging and stimulation, kinematic measures of large-scale body movements, and measures of sensorimotor synchronization at the millisecond time scale). These basic social and cognitive processes can be generalized to similar, not necessarily musical, cognitive phenomena. In fact, the human capacity for music might have evolved as a tool that fosters social bonding and group cohesion in general (Kirschner and Tomasello 2010).

Group level musical coordination can be considered as a microcosm of social interaction, where individuals function as processing units embedded within a complex system (i.e. the ensemble) whose goal entails aesthetic and emotional communication. Each unit possesses the capability to transmit sensory information non-verbally, as well to decode other's movement via the mirror matching system. As information flows along these two channels simultaneously, each unit—and the system as a whole—relies upon predictive models to meet the real-time demands of interpersonal coordination. Thus, the musical ensemble constitutes a *dynamical* system that possesses important constraints due to the fact that co-performers behave in a complex but formalized (rule-based) manner dictated by musical conventions and often a notated score. We argue that these constraints are beneficial from an experimental perspective. The score played by the musicians, the inherently rewarding experience of achieving interpersonal synchrony, and the intuitive and natural form of social interaction collectively translate into ready-made experimental tasks, high levels of intrinsic motivation, and rich ecological settings.

11.5   The Case of Orchestras and Quartets

In a series of studies, we measured sensorimotor communication via the Granger algorithm, applied to movement orchestras and quartets kinematics. The Granger methodology tests whether knowledge of one signal significantly increase prediction accuracy of the future state of another signal (Granger 1969). In other terms, the algorithm measures whether one musician behavior has any influence on the behavior of another musician.

Our first study specifically focused on the communication dynamics between orchestra conductors and the musicians (D'Ausilio et al. 2012). We described the complex pattern of sensorimotor communication in the orchestra scenario as well as the effect of such interaction on the perceived quality of the musical output. We showed that the driving-force strengths towards musicians and communication strength among players was modulated across pieces. Interestingly, when the increased influence of the conductor was paralleled by a significant reduction in inter-musician communication, the musical piece was better rated by independent professional musicians. Rather, when the conductor exerted an increased drive, while this was not paralleled by a reduction in inter-musician influences, resulted in worse musical performance. We concluded that a simple increase in conduction drive might be detrimental to perceived quality if this is not followed by a reduced inter-musician interaction (Fig. 11.2).

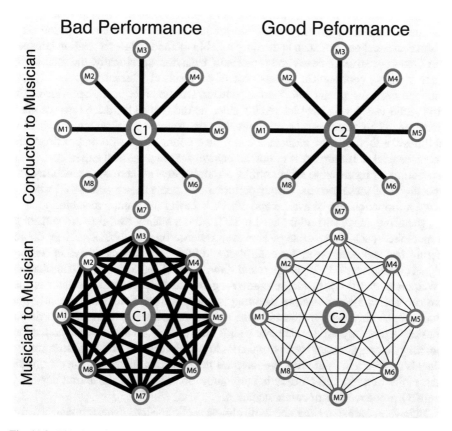

Fig. 11.2 This figure shows a graphical description of the main results obtained by our 2012 study (D'Ausilio et al. 2012). Here we related the two indexes derived by the application of the Granger algorithm to movement kinematics—"Conductor to Musicians" and "Musicians to Musicians"— and quality of performance. Quality of performance was provided by expert musician on several dimensions including technical and esthetic aspects. Results shows that good performance was associated to large "Conductor to Musicians" drive paralleled by a reduction in "Musicians to Musicians" communication

The second study, on the behavioral coordination in quartets, showed that information flow between all musicians as a group, was larger during segments characterized by technical coordination difficulty. These effects corroborate the hypothesis that musicians coordinate behavior by maximizing their coordinative efforts in specific and critical moments in time. Furthermore, we could show the emerging of an implicit hierarchy between musicians, with the first violin being most frequently the leader of the group. Then, by applying a perturbation to the communication pattern between the first violin and the rest of the quartet, we evidenced specific changes in group sensorimotor communication. Inter-musician communication increased whereas the driving force originated from the first violin (the only one knowing when and what alteration to perform) decreased.

The perturbation forces an artificial and temporally confined change in the group dynamics. Such an approach, although it introduces a rather unnatural condition to the normal musical performance is the only reliable method to infer a relation between our Granger-derived indexes and musicians' behavior. Specifically, the increase in inter-musician communication was probably a result of increased uncertainty and thus the need for greater information transfer. On the other hand, behavior of the first violin was particularly interesting since he did reduce his drive towards other musicians. This result can be interpreted in the framework of action anticipation abilities. In fact, quartet musicians can derive coordinative information from the score, as well as from models of others' behavior. Strong musical expertise, as well as continuous rehearsal as an ensemble, certainly helps in increasing reliability and specificity of these models. In our perturbation approach, we suddenly make the score information almost useless and other's behavior more unpredictable.

Therefore, musicians might need to shift from a mainly feed-forward control to a more feed-back based strategy. However, reliance on a feed-back strategy do not permit fast and accurate motor control due to inherent temporal delays in sensory signal (Wolpert et al. 1995; Wolpert and Kawato 1998) and multi-agent coordination (Wolpert et al. 2003). In fact, the advantage of modeling other's behavior is that we can anticipate and sample incoming sensory information less frequently and mainly to confirm our hypotheses on the environment. Therefore, in this experiment musicians probably had to completely shift their normal manner of communication, possibly "reading" others' behavioral cues in real-time while "sending" informative signals more often. Therefore, we interpret the reduction in driving force from the first violin as a reduced efficacy in communication caused by the abrupt departure from a learned manner of communication.

Behavior complexity, together with almost negligible lags in performance (Kokal and Keysers 2010) suggests that purely reactive mechanisms cannot be effective. Others' action anticipation is a necessary prerequisite for successful joint action control (Knoblich and Jordan 2003; Pecenka and Keller 2011). Generally speaking, expertise has been shown to support other's action classification by modeling future information earlier (Aglioti et al. 2008) and with greater detail (Calvo-Merino et al. 2005). Our musicians were almost at the top level of musical skills, thus suggesting that a strong degree of anticipation of others' action was taking place.

Beyond global descriptions of musician's pattern of relationships, the complexity of these kinds of scenario could also be exploited to distinguish and evaluate the existence of multiple channels of communication as well as their respective role in efficient coordination. In previous studies, one representative kinematic parameter was used to extract global coordination (D'Ausilio et al. 2012; Badino et al. 2014). However, we know that movements of different body parts may convey substantially different types of information. For instance, bow movements in violinists directly control the sound output (i.e., instrumental gestures), whereas complementary torso oscillations may serve a secondary communicative purpose (ancillary gestures; D'Ausilio et al. 2015b). More importantly, movements of different body parts may act as different channels of communication, possibly with different roles depending on the specific communication mode. For example, within a quartet, musicians have specific roles while in orchestras, musicians generally play in distinct sections (e.g. sections of violinists). This means that in the orchestra scenario, different modes of communication coexist: a complementary coordination with the conductor and other musicians, in parallel with an imitative coordination with musicians of the same group (playing the same score).

In a recent study (Hilt et al. 2019), we demonstrated that the pattern of sensori-motor information carried by two selected movements (head and bow) are distinct. Bow kinematics exhibit a robust leader-follower relationship between the conductor and the two violinists' sections. This pattern is robust and is not affected by the exper-imental manipulation of the sensorimotor information flow (perturbed condition) except for a decrease in communication between the first section and the conductor. Perturbation consisted in a 180 degrees' rotation of the first line of violinist, such that the conductor has direct visual contact with the second line only. The fact that the perturbation did not dramatically alter the information exchanged via instrumental movements suggests an important role of memory, score reading and residual sensory cues. Ancillary movements, instead, are supposed to convey slower frequency signals possibly related to the expressive component of musical execution, which is more likely to be affected by perturbation of the interaction dynamics. In fact, in head data, the perturbation produced clear alteration of the communication pattern. Communi-cation between the first section and the conductor or the second section was reduced. At the same time, communication between the second section and the conductor increased in both directions. This global increase suggests a greater need for infor-mation exchange during the perturbation. Instead, moving to the relationship between first and second line, we observe a complete reversal of their mutual communication. Before the perturbation, the first section provided larger causal drive towards the second, while after, the second section lead the first. During the perturbation, the first section no longer had visual contact with the conductor, significantly reducing his role in leading orchestra dynamics. Although violinists of the second section did not actually change their position, they seem to increase their normal communica-tion with conductors, while at the same time they dramatically change the way the communicate with S1 (Fig. 11.3).

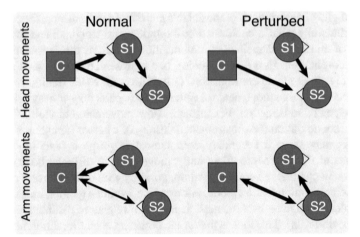

Fig. 11.3 Graphical description of the results obtained in a study investigating hand and arm movements in orchestras. Here we manipulated the information flow between the conductor, the first and the second line of violinists. The perturbation was a half-turn rotation of the first section, thus facing the second line. Perturbation to this system produces a reconfiguration of the sensorimotor communication network leading to a new centrality of the second line. Interestingly, effects were mostly limited to ancillary movements, which are supposed to convey the expressive component of musical performance

11.6   Concluding Remarks

In conclusions, the present studies show independent confirmations that Granger methods on movement kinematics measure information flow between complex multi-agent interactions. Measuring the temporal deployment of sensori-motor communication between several individuals, in a realistic scenario, opens to a series of important applications. In fact, current communication analyses hardly offer quantitative results when dealing with complex natural situations. At the same time, our data hint to the fact that efficient sensori-motor communication, may not reside in the unspecific increase in information transfer. One intriguing possibility we suggest is that the fabric of expert interaction is not the ability to communicate per se, but rather the ability to modulate such information transfer when this is more necessary.

Critically, this line of research has merely scratched the surface of one of the most underexplored issue in neuroscience. In fact, our data suggest that sensorimotor communication travels through multiple channels, each with a potentially different bandwidth and serving different purposes during behavioral coordination. If on one hand we know that informational exchange might happen at multiple time scales (i.e. instrumental and ancillary movements), on the other we do not know exactly which kind of information and how it is channeled through body movements, the degree of awareness of these exchanges, the role played by expertise and most importantly its effect coordination efficacy. These are the fundamental questions that, when answered, will move our understanding from an incomplete and local description of a complex phenomenon to a real grasp on how groups of individuals manage to gather together to become one entity.

Addendum

The Phenomenology of Musical Motricity: Perspectives from an Orchestra Conductor (by Serâ Tokay)

Own-Body, Kinesthesia and Music

I have always considered particularly useful the phenomenological approach to examine the experience of the musician interpreting the works of the classical repertoire (Tokay, 2016). The instrumentalist's sensorimotor experience is the "place" where it is possible to bring together the muscular kinaesthesia of the instrumental gesture, the reactivation of the creative intuition of the composer and the empathic resonance resulting from what other musicians and the music-lovers hear. As musician and philosopher, my approach makes use of a theory of the constitutive function of the kinaesthetic system derived from the work of Husserl (1966, 1982), re-oriented, in the direction of a theory of the constitution of the musical sound, insofar as the latter emanates from the sensorimotor activity of the musician. In the following sections, inspired by Husserl's perspective, I will start discussing the empathic process linking conductor, musicians and audience, I will then continue along the theory of an active production, controlled by the technical savoir-faire of the musicians and by the psychophysical bidirectional link established by the conductor with them. Finally, I will examine the experimental results presented in this chapter from a perspective centered on a theory of the aesthetic object (the musical object insofar as it responds to the artistic norms of harmony). My approach has been developed in opposition to the physicalist reduction of the aesthetic dimension of the musical sound through its analysis in terms of physical properties of acoustic oscillations (von Helmholtz 1990; Xenakis 1971; Boulez 1987).

Empathy and Motor Entrainment

As 'Einfühlung' Lipps (1903) defined the lively feeling of being-self which an observer is able to derive from the aesthetic contemplation of another being in movement. He emphasized the direct way in which, without any intermediary judgment, a human being can get immersed in another interior life via observation of the facial and corporeal expressions of the other. Husserl noted that the perceptual recognition of the other requires an appreciation of his body as own body, and he amplified Einfüh-lung into a dynamic process of co-constitution, the constitution of a common world across the intentions and actions of the self and of others. In the same perspective as the co-constitution of a life-world, but by withdrawing from any transcendental abstraction into a more concrete and bodily field, I take account of those participating in a concert. The movements of the instrumentalists, all of whom are oriented towards the same musical "ideas", frame the horizon within which the sounds produced take on the value of musical sounds for those in the audience.

Conducting an Orchestra: Coordination or Anticipation?

Here I would like to stress that the integration of the actions of the instrumentalists engaged in playing a musical work depends upon the conductor's power of antici-pation. This anticipation depends upon a motor apprehension of musical time. The result of a quite special technical mastery of the movement of the hands and arms. The speed with which neuromuscular innervation is produced enables the conductor to elicit a haptic analogue of the motor intentions of the instrumentalist. As for the musicians, this anticipation of the kinaesthetic temporality of their own movements by those of the conductor is necessary if one is to grasp the expressive intentions guiding the hands, the gaze, the breath and the bodily posture of the conductor. Without all this it is impossible to understand the affective character of the work, its temporality and its rhythm. According to this view, conducting does not mean to create a hierarchical subordination between the micro-intentions of the instrumental-ists and the directional intentions of the conductor. Conversely, there is a continuous alternation between the common goal (harmony) and the singular goals of individual musicians, regulated by a collective desire to play the work as well as possible.

As an example I would like to remember the very start of the 4th Symphony of Schumann, which begins with an A without rhythm in *tutti*, where a rhythm in ¾ has to be conducted in 6 times interspersed with silences, and in an atmosphere agitated by a continual alternation between forte-piano and crescendo-diminuendo, against a background of *rubato*. In this context the entire orchestra shares a single *telos* to intuit a *tempo*. In such a tempestuous climate each musician has to count his quavers individually, even while sharing the same sense of time, whose unity is established by an empathic relation with the movements of the conductor.

Discussion of Some Experimental Results

1. *The Driving Force of the Conductor*

In this work we have analyzed the kinematic recordings of violin players belonging to a chamber orchestra while playing pieces of Mozart under the direction of two different conductors. As *Driving Force*, we considered the amount of causality between conductor's hand and baton velocities and violinists' bows determined according to Granger's method. This concept of causality depends on the idea that when one has two variables which are relatively independent (like the respective trajectories of the hands of the musicians and of the conductor), but which depend upon each other in such a way that knowledge of the past values of one trajectory make it possible to predict the value of the other (the conductor getting ahead of the musicians), the first 'causes' the second (in Granger's sense of that word). The comparison of the curves described by a recording of the conductors' batons and the bows of the violinists show that the invited conductor (who directs in advance) exerts a driving force over the musicians that is both more frequent and stronger than that of the principal conductor. A comparison of the curves recording the bows of the violinists when they play under the direction of the 2 conductors shows that the driving force exerted by the musicians on each other is that much weaker when the driving force of the conductor is stronger and when he figures as the source of the unity of the ensemble. It is plausible that the measure of the conductor's driving force represent an acceptable quantification of the intersubjective experience that we describe from a phenomenological point of view with the concept of motor empathy acting between the conductor and the musicians.

2. *The leadership of the conductor as soft entrainment*

In an orchestra, the leadership exerted by the conductor consists in getting his musicians to play together. So it should be possible to measure the efficiency of the conductor by the degree of entrainment he instils in the orchestra. But unlike the metronome, which achieves a perfect synchronization without any musical value, the entrainment of the orchestra owes its aesthetic character to a subtle modulation of the dynamics of the entrainment (soft entrainment). The emergence of a collective rhythm on the basis of the individual rhythms of the musicians considered as coupled oscillators is made possible by a phase synchronization in the oscillations, which is mathematically described by the Kuramoto model (Kuramoto 1984). An index of synchronization measuring the soft entrainment of the orchestra is calculated for each measure, and for each conductor. This index, coupled with an evaluation of the artistic quality of the execution, offers a criterion for the efficiency of the leadership of the conductor, one that confirms the score already obtained. The most efficient conductor is the one who exerts the most considerable influence on the musicians. The convergence of the results obtained with the same participants in applying two different mathematical models to the data recorded, the one issuing from econometrics (Granger causality), the other from the analytic mechanics of coupled oscillators (Kuramoto model) proves that anticipation is essential to the direction of an orchestra, whatever the theoretical model employed.

3. *Extended empathy through ancillary movements*

This experiment has looked at significant differences between the ancillary movements of the head of the musicians when the musicians benefit from a normal vision of the movements of the conductor, and when part of them (the first row) are turned by 180°, and can only obtain indirect information on tempo, upbeat, bow movements, nuances, beats, fermata, etc. through the visual observation of the gestures of musicians of the second row that does enjoy a direct vision of the conductor. The recordings in question have been made with the overture to Rossini's Opera: *Il Signor Bruschino*. Results were clear: the musicians who are unable to see the conductor make more ancillary movements of the head than when they are able to see her/him. This difference is a measurable characteristic of the movement, a geometrical precursor of the intentional orientation of this movement, indicative of the effort the musicians are obliged to make to mitigate with ancillary movements their less than optimal condition. They are able to compensate for the lack of vision of the conductor with glances at the instrumental movements of their colleagues in the other group facing them. If they nevertheless still remain ensemble, it is only because all the musical information provided by their conductor has been relayed to them by the instrumental movements of this other group of musicians. And so it is that a common behavioural pattern emerges, one that constitutes an inter-subjective field of intentions expressed through gestures encompassing the musicians of both groups. Here, the experimental manipulation achieves an exceptional extension of the inter-subjective space of communication, putting to the test the conductor's ability to convey his artistic intentions no matter how great the distance to be crossed.

What Can We Learn from These Investigations?

In summary, and from my perspective, these are the main outcome of this investigation:

(1) In the body of the musician, from which every affectively qualified action proceeds, converge both the experiences of the interpreter and of the auditor.
(2) Only a conductor capable of an auto-affective anticipation finds herself in a position to promote the emotive and motivating principle necessary for the musician to play well.
(3) An orchestra confronting its conductor awaits from her a non-verbal motor communication, which in turn depends upon her capacity to surpass the initially inevitable affective confrontation with the orchestra by being ahead of herself through a perfectly mastered gestural strategy.
(4) As a function of its bodily mastery, our understanding of musical experience requires both a subjective reflexivity of a phenomenological order and an objectivity derived from a neuroscientific approach.

References

Aglioti, S. M., Cesari, P., Romani, M., & Urgesi, C. (2008). Action anticipation and motor resonance in elite basketball players. *Nature Neuroscience, 11*(9), 1109–1116.

Avenanti, A., Bolognini, N., Maravita, A., & Aglioti, S. M. (2007). Somatic and motor components of action simulation. *Current Biology: CB, 17*(24), 2129–2135.

Babiloni, C., Buffo, P., Vecchio, F., Marzano, N., Del Percio, C., Spada, D., et al. (2012). Brains "in concert": Frontal oscillatory alpha rhythms and empathy in professional musicians. *NeuroImage, 60*(1), 105–116.

Badino, L., D'Ausilio, A., Glowinski, D., Camurri, A., & Fadiga, L. (2014). Sensorimotor communication in professional quartets. *Neuropsychologia, 55,* 98–104.

Bangert, M., & Altenmüller, E. O. (2003). Mapping perception to action in piano practice: A longitudinal DC-EEG study. *BMC Neuroscience, 4,* 26.

Bangert, M., Peschel, T., Schlaug, G., Rotte, M., Drescher, D., Hinrichs, H., et al. (2006). Shared networks for auditory and motor processing in professional pianists: Evidence from fMRI conjunction. *NeuroImage, 30*(3), 917–926.

Borra, E., Belmalih, A., Calzavara, R., Gerbella, M., Murata, A., Rozzi, S., & Luppino, G. (2008). Cortical connections of the macaque anterior intraparietal (AIP) area. *Cerebral Cortex, 18*(5):1094–1111. (New York, NY: 1991).

Borroni, P., Montagna, M., Cerri, G., & Baldissera, F. (2005). Cyclic time course of motor excitability modulation during the observation of a cyclic hand movement. *Brain Research, 1065*(1–2), 115–124. https://doi.org/10.1016/j.brainres.2005.10.034.

Boulez, P. (1987). *Penser la musique aujourd'hui.* Paris: Gallimard.

Buccino, G., Binkofski, F., Fink, G. R., Fadiga, L., Fogassi, L., Gallese, V., et al. (2001). Action observation activates premotor and parietal areas in a somatotopic manner: An fMRI study. *The European Journal of Neuroscience, 13*(2), 400–404.

Calvo-Merino, B., Glaser, D. E., Grezes, J., Passingham, R. E., & Haggard, P. (2005). Action observation and acquired motor skills: An FMRI study with expert dancers. *Cerebral Cortex, 15*(8), 1243–1249. (New York, NY: 1991).

Candidi, M., Sacheli, L. M., Mega, I., & Aglioti, S. M. (2014). Somatotopic mapping of piano fingering errors in sensorimotor experts: TMS studies in pianists and visually trained musically naives. *Cerebral Cortex, 24*(2), 435–443. (New York, NY: 1991).

Cicchini, G. M., Arrighi, R., Cecchetti, L., Giusti, M., & Burr, D. C. (2012). Optimal encoding of interval timing in expert percussionists. *The Journal of Neuroscience: The Official Journal of the Society for Neuroscience, 32*(3), 1056–1060.

Clark, A., & Grush, R. (1999). Towards a cognitive robotics. *Adaptive Behavior, 7,* 5–16.

Couzin, I. D., Krause, J., Franks, N. R., & Levin, S. A. (2005). Effective leadership and decision-making in animal groups on the move. *Nature, 433*(7025), 513–516.

D'Ausilio, A., Bartoli, E., & Maffongelli, L. (2015a). Grasping synergies: A motor-control approach to the mirror neuron mechanism. *Physics of life Reviews, 12,* 91–103.

D'Ausilio, A., Brunetti, R., Delogu, F., Santonico, C., & Belardinelli, M. O. (2010). How and when auditory action effects impair motor performance. *Experimental Brain Research, 201*(2), 323–330.

D'Ausilio, A., Novembre, G., Fadiga, L., & Keller, P. E. (2015b). What can music tell us about social interaction? *Trends in Cognitive Sciences, 19*(3), 111–114.

D'Ausilio, A., Altenmuller, E., Olivetti Belardinelli, M., & Lotze, M. (2006). Cross-modal plasticity of the motor cortex while listening to a rehearsed musical piece. *The European Journal of Neuroscience, 24*(3), 955–958.

D'Ausilio, A., Badino, L., Li, Y., Tokay, S., Craighero, L., Canto, R., et al. (2012). Leadership in orchestra emerges from the causal relationships of movement kinematics. *PLoS ONE, 7*(5), e35757.

Decety, J., Grezes, J., Costes, N., Perani, D., Jeannerod, M., Procyk, E., et al. (1997). Brain activity during observation of actions. Influence of action content and subject's strategy. *Brain: A Journal of Neurology, 120*(Pt 10), 1763–1777.

Elbert, T., Pantev, C., Wienbruch, C., Rockstroh, B., & Taub, E. (1995). Increased cortical representation of the fingers of the left hand in string players. *Science (New York, NY), 270*(5234), 305–307.

Engel, A., Bangert, M., Horbank, D., Hijmans, B. S., Wilkens, K., Keller, P. E., et al. (2012). Learning piano melodies in visuo-motor or audio-motor training conditions and the neural correlates of their cross-modal transfer. *NeuroImage, 63*(2), 966–978.

Fadiga, L., Fogassi, L., Pavesi, G., & Rizzolatti, G. (1995). Motor facilitation during action observation: A magnetic stimulation study. *Journal of Neurophysiology, 73*(6), 2608–2611.

Fairhurst, M. T., Janata, P., & Keller, P. E. (2013). Being and feeling in sync with an adaptive virtual partner: Brain mechanisms underlying dynamic cooperativity. *Cerebral Cortex, 23*(11), 2592–2600. (New York, NY: 1991).

Fairhurst, M. T., Janata, P., & Keller, P. E. (2014). Leading the follower: An fMRI investigation of dynamic cooperativity and leader-follower strategies in synchronization with an adaptive virtual partner. *NeuroImage, 84,* 688–697.

Fazio, P., Cantagallo, A., Craighero, L., D'Ausilio, A., Roy, A. C., Pozzo, T., et al. (2009). Encoding of human action in Broca's area. *Brain: A Journal of Neurology, 132*(Pt 7), 1980–1988. https://doi.org/10.1093/brain/awp118.

Fogassi, L., Ferrari, P. F., Gesierich, B., Rozzi, S., Chersi, F., & Rizzolatti, G. (2005). Parietal lobe: From action organization to intention understanding. *Science (New York, NY), 308*(5722), 662–667.

Friston, K., Mattout, J., & Kilner, J. (2011). Action understanding and active inference. *Biological cybernetics, 104,* 137–160. https://doi.org/10.1007/s00422-011-0424-z.

Frith, C. D. (2008). Social cognition. *Philosophical Transactions of the Royal Society of London. Series B, Biological sciences, 363*(1499), 2033–2039.

Gallese, V., Fadiga, L., Fogassi, L., & Rizzolatti, G. (1996). Action recognition in the premotor cortex. *Brain: A Journal of Neurology, 119*(Pt 2), 593–609.

Gazzola, V., & Keysers, C. (2009). The observation and execution of actions share motor and somatosensory voxels in all tested subjects: Single-subject analyses of unsmoothed fMRI data. *Cerebral Cortex, 19*(6), 1239–1255. (New York, NY: 1991).

Glowinski, D., Mancini, M., Cowie, R., Camurri, A., Chiorri, C., & Doherty, C. (2013). The movements made by performers in a skilled quartet: A distinctive pattern, and the function that it serves. *Frontiers in Psychology, 4,* 841.

Grafton, S. T., & Hamilton, A. F. (2007). Evidence for a distributed hierarchy of action representation in the brain. *Human Movement Science, 26*(4), 590–616.

Granger, C. W. J. (1969). Investigating causal relations by econometric models and cross-spectral methods. *Econometrica., 37,* 424–438.

Grezes, J., & Decety, J. (2001). Functional anatomy of execution, mental simulation, observation, and verb generation of actions: A meta-analysis. *Human Brain Mapping, 12*(1), 1–19.

Hari, R., Forss, N., Avikainen, S., Kirveskari, E., Salenius, S., & Rizzolatti, G. (1998). Activation of human primary motor cortex during action observation: A neuromagnetic study. *Proceedings of the National Academy of Sciences of the United States of America, 95*(25), 15061–15065.

Hari, R., Henriksson, L., Malinen, S., & Parkkonen, L. (2015). Centrality of social interaction in human brain function. *Neuron, 88*(1), 181–193.

Haslinger, B., Erhard, P., Altenmuller, E., Schroeder, U., Boecker, H., & Ceballos-Baumann, A. O. (2005). Transmodal sensorimotor networks during action observation in professional pianists. *Journal of Cognitive Neuroscience, 17*(2), 282–293.

Haueisen, J., & Knösche, T. R. (2001). Involuntary motor activity in pianists evoked by music perception. *Journal of Cognitive Neuroscience, 13*(6), 786–792.

Hilt, P. M., Badino, L., D'Ausilio, A., Volpe, G., Fadiga, L., & Camurri, A. (2019). Multi-layer adaptation of group coordination in musical ensembles. *Scientific Reports, 9,* 5854.

Hurley, S. (2008). The shared circuits model (SCM): How control, mirroring, and simulation can enable imitation, deliberation, and mindreading. *The Behavioral and Brain Sciences, 31*(1), 1–22.

Husserl, E. (1966). Zur Phänomenologie des Inneren Zeitbewusstseins (1893–1917). In V. R. Boehm (Ed.), *Husserliana X*. La Haye: Martinus Nijhoff.

Husserl, E. (1982). *Idées directrices pour une phénoménologie et une philosophie phénoménologique pures*, vol. II, *Recherches phénoménologiques pour la constitution*, trad. E. Escoubas. Paris: PUF.

Iacoboni, M., Molnar-Szakacs, I., Gallese, V., Buccino, G., Mazziotta, J. C., & Rizzolatti, G. (2005). Grasping the intentions of others with one's own mirror neuron system. *PLoS Biology, 3*(3), e79.

Iacoboni, M., Woods, R. P., Brass, M., Bekkering, H., Mazziotta, J. C., & Rizzolatti, G. (1999). Cortical mechanisms of human imitation. *Science (New York, NY), 286*(5449), 2526–2528.

Kilner, J. M., Neal, A., Weiskopf, N., Friston, K. J., & Frith, C. D. (2009). Evidence of mirror neurons in human inferior frontal gyrus. *The Journal of Neuroscience: The Official Journal of the Society for Neuroscience, 29*(32), 10153–10159.

Kirschner, S., & Tomasello, M. (2010). Joint music making promotes prosocial behavior in 4-year-old children. *Evolution and Human Behavior, 31*(5), 354–364.

Knoblich, G., & Flach, R. (2001). Predicting the effects of actions: Interactions of perception and action. *Psychological Science, 12*(6), 467–472.

Knoblich, G., & Jordan, J. S. (2003). Action coordination in groups and individuals: Learning anticipatory control. *Journal of Experimental Psychology: Learning, Memory, and Cognition, 29*(5), 1006–1016.

Kokal, I., Engel, A., Kirschner, S., & Keysers, C. (2011). Synchronized drumming enhances activity in the caudate and facilitates prosocial commitment—If the rhythm comes easily. *PLoS ONE, 6*(11), e27272.

Kokal, I., & Keysers, C. (2010). Granger causality mapping during joint actions reveals evidence for forward models that could overcome sensory-motor delays. *PLoS ONE, 5*(10), e13507.

Kuramoto, Y. (1984). *Chemical oscillations, waves, and turbulence*. New York, NY, USA: Springer.

Lahav, A., Saltzman, E., & Schlaug, G. (2007). Action representation of sound: Audiomotor recognition network while listening to newly acquired actions. *The Journal of Neuroscience: The Official Journal of the Society for Neuroscience, 27*(2), 308–314.

Limb, C. J., & Braun, A. R. (2008). Neural substrates of spontaneous musical performance: An FMRI study of jazz improvisation. *PLoS ONE, 3*(2), e1679.

Lipps, T. (1903). Einfühlung, innere Nachahmung, und Organempfindung. *Archiv für die Gesamte Psychologie, 1*(2), 185–204.

Loehr, J. D., Kourtis, D., Vesper, C., Sebanz, N., & Knoblich, G. (2013). Monitoring individual and joint action outcomes in duet music performance. *Journal of Cognitive Neuroscience, 25*(7), 1049–1061.

Lotze, M., Scheler, G., Tan, H. R., Braun, C., & Birbaumer, N. (2003). The musician's brain: Functional imaging of amateurs and professionals during performance and imagery. *NeuroImage, 20*(3), 1817–1829.

Luppino, G., Murata, A., Govoni, P., & Matelli, M. (1999). Largely segregated parietofrontal connections linking rostral intraparietal cortex (areas AIP and VIP) and the ventral premotor cortex (areas F5 and F4). *Experimental Brain Research, 128*(1–2), 181–187.

Murata, A., Fadiga, L., Fogassi, L., Gallese, V., Raos, V., & Rizzolatti, G. (1997). Object representation in the ventral premotor cortex (area F5) of the monkey. *Journal of Neurophysiology, 78*(4), 2226–2230.

Nagy, M., Akos, Z., Biro, D., & Vicsek, T. (2010). Hierarchical group dynamics in pigeon flocks. *Nature, 464*(7290), 890–893.

Newman-Norlund, R. D., van Schie, H. T., van Zuijlen, A. M., & Bekkering, H. (2007). The mirror neuron system is more active during complementary compared with imitative action. *Nature Neuroscience, 10*(7), 817–818.

Novembre, G., Ticini, L. F., Schutz-Bosbach, S., & Keller, P. E. (2012). Distinguishing self and other in joint action. Evidence from a musical paradigm. *Cerebral Cortex, 22*(12), 2894–2903. (New York, NY: 1991).

Pantev, C., Oostenveld, R., Engelien, A., Ross, B., Roberts, L. E., & Hoke, M. (1998). Increased auditory cortical representation in musicians. *Nature, 392*(6678), 811–814.

Pecenka, N., & Keller, P. E. (2011). The role of temporal prediction abilities in interpersonal sensorimotor synchronization. *Experimental Brain Research, 211*(3–4), 505–515.

Petrides, M., & Pandya, D. N. (2009). Distinct parietal and temporal pathways to the homologues of Broca's area in the monkey. *PLoS Biology, 7*(8), e1000170.

Ragert, M., Schroeder, T., & Keller, P. E. (2013). Knowing too little or too much: The effects of familiarity with a co-performer's part on interpersonal coordination in musical ensembles. *Frontiers in Psychology, 4,* 368.

Rands, S. A., Cowlishaw, G., Pettifor, R. A., Rowcliffe, J. M., & Johnstone, R. A. (2003). Spontaneous emergence of leaders and followers in foraging pairs. *Nature, 423*(6938), 432–434.

Rizzolatti, G., & Craighero, L. (2004). The mirror-neuron system. *Annual Review of Neuroscience, 27,* 169–192.

Rizzolatti, G., Fadiga, L., Matelli, M., Bettinardi, V., Paulesu, E., Perani, D., et al. (1996). Localization of grasp representations in humans by PET: 1. Observation versus execution. *Experimental Brain Research, 111*(2), 246–252.

Rizzolatti, G., & Luppino, G. (2001). The cortical motor system. *Neuron, 31*(6), 889–901.

Rizzolatti, G., & Sinigaglia, C. (2010). The functional role of the parieto-frontal mirror circuit: Interpretations and misinterpretations. *Nature Reviews Neuroscience, 11*(4), 264–274.

Rozzi, S., Calzavara, R., Belmalih, A., Borra, E., Gregoriou, G. G., Matelli, M., & Luppino, G. (2006). Cortical connections of the inferior parietal cortical convexity of the macaque monkey. *Cerebral Cortex, 16*(10), 1389–1417. (New York, NY: 1991).

Sänger, J., Muller, V., & Lindenberger, U. (2012). Intra- and interbrain synchronization and network properties when playing guitar in duets. *Frontiers in Human Neuroscience, 6,* 312.

Sänger, J., Muller, V., & Lindenberger, U. (2013). Directionality in hyperbrain networks discriminates between leaders and followers in guitar duets. *Frontiers in Human Neuroscience, 7,* 234.

Schilbach, L., Timmermans, B., Reddy, V., Costall, A., Bente, G., Schlicht, T., et al. (2013). Toward a second-person neuroscience. *The Behavioral and Brain Sciences, 36*(4), 393–414.

Schlaug, G., Jancke, L., Huang, Y., & Steinmetz, H. (1995). In vivo evidence of structural brain asymmetry in musicians. *Science (New York, NY), 267*(5198), 699–701.

Sebanz, N., Bekkering, H., & Knoblich, G. (2006). Joint action: Bodies and minds moving together. *Trends in cognitive sciences, 10*(2), 70–76.

Sommerville, J. A., & Decety, J. (2006). Weaving the fabric of social interaction: Articulating developmental psychology and cognitive neuroscience in the domain of motor cognition. *Psychonomic Bulletin & Review, 13*(2), 179–200.

Tokay, S. (2016). *Le corps musicien.* Montréal: Editions Liber.

Vernon, D., Von Hofsten, C., & Fadiga, L. (2011). *A roadmap for cognitive development in humanoid robots.* Berlin: Springer.

von Helmholtz, H. (1990). *Théorie physiologique de la musique.* Paris: Jacques Gabay.

Wilson, M., & Knoblich, G. (2005). The case for motor involvement in perceiving conspecifics. *Psychological Bulletin, 131*(3), 460–473.

Wing, A. M., Endo, S., Bradbury, A., & Vorberg, D. (2014). Optimal feedback correction in string quartet synchronization. *Journal of the Royal Society, Interface, 11*(93), 20131125.

Wolpert, D. M., Doya, K., & Kawato, M. (2003). A unifying computational framework for motor control and social interaction. *Philosophical Transactions of the Royal Society of London. Series B, Biological sciences, 358*(1431), 593–602.

Wolpert, D. M., Ghahramani, Z., & Jordan, M. I. (1995). An internal model for sensorimotor integration. *Science (New York, NY), 269*(5232), 1880–1882.

Wolpert, D. M., & Kawato, M. (1998). Multiple paired forward and inverse models for motor control. *Neural Networks: The Official Journal of the International Neural Network Society, 11*(7–8), 1317–1329.

Xenakis, I. (1971). *Musique architecture*. Paris: Casterman.

Zatorre, R. J., Chen, J. L., & Penhune, V. B. (2007). When the brain plays music: Auditory-motor interactions in music perception and production. *Nature Reviews Neuroscience, 8*(7), 547–558.

Part IV
The Arts: Digital Arts

Chapter 12
Artistic Practices in Digital Space: An Art of the Geometries of Movement?

François Garnier

Abstract Tamar Flash, Alain Berthoz, and Daniel Bennequin have suggested that we, humans, are using different cognitive geometrical strategies to perceive space depending on the type of environment, the action we perform, and the spatial distance of its realization with respect to the observer. Various brain networks using different geometries are involved in these different action spaces. Therefore this type of perceptual structuring of our environment should also influence the way we represent in arts the spaces that surround us. Here we hypothesize that it should be possible to find structural analogies between these different perceptual geometric strategies and the modes of representation used in different visual arts such as in painting, sculpture, traditional and stereoscopic cinema, and new forms of spatial mediation, immersive spaces, and virtual reality. We give examples of pictorial representations, which suggest that that the notion of "formats" in visual arts, which are still used by creators, could be explained by their capacity to express the different dimensions of expression of the "geometries of movement."

12.1 Introduction

In their recent works, Tamar Flash, Alain Berthoz, and Daniel Bennequin[1] submit that we use different geometric cognitive strategies to perceive our environment and act upon it depending on the type of action we engage in and the spatial distance of its realization.

This type of perceptive structuration of our environment should influence the way we represent the spaces that surround us. As a practice aiming to reproduce or

---

[1] Bennequin et al. (2009).

Translated from the French by Jennifer Gay.

F. Garnier (✉)
Ensad, PSL University, Ensad Lab, Spatail Media, Paris, France
e-mail: francois.garnier@ensad.fr

© Springer Nature Switzerland AG 2021
T. Flash and A. Berthoz (eds.), *Space-Time Geometries for Motion and Perception in the Brain and the Arts*, Lecture Notes in Morphogenesis,
https://doi.org/10.1007/978-3-030-57227-3_12

transcribe our relationship to the world, even to appropriate it, art should be influenced by these cognitive processes. In this article, we hypothesize that it should be possible to find structural analogies between these geometric perceptual strategies and the modes of representation used in art and communication.

First, on the basis of these articles and discussions with their authors, we will attempt to show how these perceptive spaces structure our environment according to the distance of considered actions.

We will then briefly present various visual arts practices in an attempt to see which representation strategies have been employed by artists to share similar actions: in the plastic arts, painting and sculpture, in traditional and stereoscopic cinema, and in new forms of spatial mediation such as immersive spaces and virtual reality.

It is not our intention to prove or demonstrate anything in particular, but rather to propose (or initiate) an open consideration—subjective and perfectible, of course—in view of developing hypotheses.

12.2 Structure of Perceptive Spaces

By basing ourselves on the descriptions provided by Alain Berthoz and Daniel Bennequin in their article,[2] we will attempt to schematically represent the spaces of action/perception that structure our environment.

We are aware that this representation by distance is reductive and that in movement, time is inseparable from space, both in its prediction and in its realization.

These areas are defined by specific geometric cognitive strategies, linked to types of actions envisaged or carried out. Their limits, if we try to describe them in distance, are therefore not fixed and must be imagined as flexible envelopes evolving in time according to the action envisaged and its context of execution. The way they fit together could be compared to a Russian doll.

This oriented choice, to approach the geometries of movement by their spatial expression, is motivated by the type of artistic expression we will study in this article, the visual arts. It would be interesting to develop a similar reflection structured by a time scale, by analyzing the arts of time such as music and living art, dance, theatre, performance…

Body Space: (From Inside the Body to 20 cm)
Body space, which is reconstructed in a 'body schema' in networks located in the temporo-parietal junction, was first shown in epileptic patients by the neurologist Wilder Penfield in Canada, who identified this brain region as responsible for 'awareness of body schema and spatial relationships'. It is known that this schema takes into account all the mechanical and dynamic properties of the real physical body, and it has also been proposed that the temporo-parietal junction contains an 'internal model' of gravity (Bennequin, Berthoz 2017).

---

[2]Bennequin and Berthoz (2017).

This first space can be considered as that of awareness of the self, of one's body, of one's otherness. It includes the feeling of the position of the body (body schema), its weight and orientation, its limits, the skin, the sensations of contact and of close proximity. This first space begins deep within the body and surrounds it to a distance of approximately 20 cm.

Prehension Space (Near Action): From 0 to 20 cm to 1 to 2 m *Near action and prehension space, which is equivalent to the space at which we can reach things with the extended hand. In this space the geometries have to include forces and dynamic properties of the objects that one manipulates or obstacles that we may encounter. Simplifying laws of movement are at work to control gestures. Actions can be made in egocentric reference frame or in object centered reference frame or, if another person is involved, in heterocentric reference frame (Bennequin, Berthoz 2017).*

This "prehension space" is superimposed on the first one (body space) and includes the entire space on which we can act without having to move our body (ambulation). It therefore includes elements outside our body, accessible objects and people with which we can potentially act or interact by extending the arm. This space allows us to change our system of reference (egocentric, heterocentric, or object-centric) by projecting ourselves into external centers of interest to feel their nature. Its default limit is that of the extended arm, but can be lengthened by using a tool. It starts at what is no longer our body, 0 to 20 cm from our skin, up to a variable limit depending on the context ranging from 1 m for the arm, and can be extended to 2 m or more by a tool.

Locomotor Space (Far Action Space): From 1 to 2 m to 10 to 20 m *Far action space, that is the space that we reach with a short locomotor trajectory (typically a room). In this space it has been shown that optimizing principles induce stereotyped trajectories. Both ego-centric and allocentric reference frames can be used as well as heterocentric ones. Evidence shows that the neural networks involved in this space are not the same as those for near action space (Bennequin, Berthoz 2017).*

Locomotor space is the visible space in which we can safely predict our movement. It is limited by what blocks our vision (walls, objects, people). Its surface is therefore heavily dependent on the nature of our environment and is limited to short and predictable trajectories, which are updated according to events. The authors indicate that, in this space, egocentric, allocentric, and heterocentric systems of reference can be used. It starts when we can no longer touch, Prehension Space (about 1 to 2 m) and stops when our senses no longer allow us to predict a safe movement. Its maximum circumference therefore varies greatly over time and depending on the context, ranging from a few meters to several dozen meters.

Environmental Space (Navigation—Imagined Motion): From 10 to 20 m to Infinity
Environmental navigation space, that cannot be explored by a short walk. Typically a city or a park that requires an allo-centric cartographic coding to be able to navigate

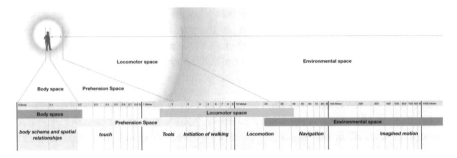

Fig. 12.1 Spatial structure of the space-time geometries

and find new paths. In addition to this modularity recent studies have identified multiple reference frames and different neural structures for 'ego-centric' (referred to an observer own body viewpoint), 'allo-centric' (map like, independent of an observer view point), or even "hetero-centric" (taking another person as a reference). This diversity of reference frame has given rise to a number of terminologies (like first or third person perspective etc.) (Bennequin, Berthoz 2017).

This is the space that is not directly accessible to us, out of physical or visual reach. It seems that we have several strategies available to us to create imagined representations of our distant environment, as well as the actions we might accomplish there (paths, models, maps, cosmogonies, etc.).

On the basis of these descriptions, it is possible to propose a first representation of the zones of perception of our environment by distance, responding to the definition of "the geometries of motion" (Fig. 12.1).

If the visual arts can be considered as paradigms aiming to reproduce or transcribe our relationship to the world, they should prove to be influenced, or even guided, by these perceptive processes. The various practices of representation in art were codified in the framework of a dialogue, an engaged sharing among artists who wanted to show actions in space and spectators who wanted to perceive them. Those uses define selections of image, space or time, modes of composition, and the codes through which artistic mediation is *expressed.*

It should therefore be possible to find structural analogies between these cognitive processes and the modes of representation used in art and communication. We propose to study the formats and codifications of use specific to different visual arts and with regard to "geometries of movement".

12.3   The "Space-Time Geometries" Theory Approach in the Visual Arts

The various practices in the visual arts can be considered as so many paradigms for representing the world. They are differentiated, among other things, by the material support whose technical characteristics they employ (stone, wood, paper, film, digital media, etc.) and by the dimensionalities of their geometric expressions (space and time). Wood can thus be both a support in graphic arts, expressed in two spatial dimensions (painting, drawing) and in sculpture, expressed in three spatial dimensions. Photographic film can be a timeless medium in photography and the support of action in cinema by integrating the temporal dimension of movement (Fig. 12.2).

We will begin by examining the older practices of the graphic arts that are expressed in two spatial dimensions (painting, drawing, photography) on a variety of two-dimensional supports (wood, stone, paper, film, digital media), and sculpture, which is expressed in three spatial dimensions on three-dimensional supports (wood, stone, digital media). Then we will study more recent practices such as the moving image, cinema, and video, which are expressed in two spatial dimensions and one time dimension, on supports such as film and digital media. And we will end with new forms of immersive mediation that are expressed in three spatial dimensions and one temporal dimension, in digital media, virtual worlds, and virtual reality. These spatial-temporal media dynamically involve the spectator's body in the act of

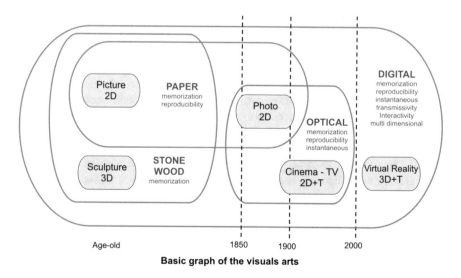

Fig. 12.2 Basic graph of visual arts

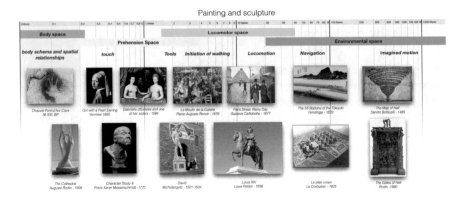

Fig. 12.3 Painting and sculpture

mediation, activating both the predictive and active potentials of the "geometries of movement".

In the Visual Arts

The visual arts include graphic arts such as drawing and painting, expressed on flat surfaces, and sculpture, expressed via three-dimensional supports. Works are generally in the form of physical objects exposed to viewers. There are two contexts in these approaches in which "geometries of movement" can be expressed, that of display—the relation between the viewer and the work—and that of the relationships among the various components that make up the work (Fig. 12.3).

In France there are three classic canvas painting formats: "figure", "paysage", and "marine" (portrait, landscape, and marine).[3]

One of the most common genres in painting[4] and sculpture[5] is the portrait, presenting the head and shoulders of the subject as a bust. The portrait is most often painted in the "figure" format, and in general presents static subjects on a realistic scale. The work is displayed in a vertical frame at eye level, viewers placing themselves face to face, at a touching distance, (prehension space) of about 1.5 m.

At a distance of about 1.5 m, or a "prehension space" type distance. The subject may be shown in the presence of an object or another subject, most often within reach. It is not uncommon that an action of touching or gazing connects the elements, opening the possibility of egocentric, heterocentric,[6] or object-centric reference. In works of this type we therefore find several of the characteristics described in "prehension space", both in the modes of display and in the intra-work relationships: distances inviting possible or actual prehension, generally static characters, and the possibility of changed systems of reference (egocentric, object-centric, and heterocentric).

---

[3] André Béguin, Dictionnaire technique de la peinture: Pour les arts, le bâtiment et l'industrie, volume 1, articles "châssis" and "organisation des surfaces", ed. A. Béguin, 2001, p. 263.

[4] *"Girl with a Pearl Earring", Johannes Vermeer, 1665.*

[5] *"Character Head 9", Franz Xaver Messerschmidt,1770.*

[6] *"Gabrielle d'Estrées and one of her sisters", unknown artist, 1594.*

Another classic painting format, the "paysage" (landscape) format, often employed horizontally, seems to correspond more specifically to the characteristics of "locomotor space". This format is generally used to paint landscapes seen from human height, inviting us to dynamically explore the space represented[7] or to immerse ourselves in active groups by identifying ourselves with the protagonists.[8]

"Locomotor space" could be associated with full-length sculpture, especially when bodies are in motion, for example walking, as in equestrian sculpture.[9] It is interesting to note that full-length sculptures are generally presented on high pedestals, placing them out of reach of viewers (spectators), and inviting them to walk around the sculpture.[10] We can also consider sculpted Greek friezes illustrating scenes of action that invite us to walk around the building, and classical gardens that invite us to stroll from sculpture to sculpture.

The "marine" format, more elongated, seems particularly adapted to "environmental space" type representations. As its name indicates, it was used, among other things, to represent vast expanses of sea and inaccessible boats, but also—in classical painting—battlefields or cities seen from a high position. However, there are many other strategies in drawing for representing "environmental space", including allocentric plan-type representations, maps, or cosmogonies.[11] In sculpture, it is models that, by playing on reductions of scale, place us in plunging and allocentric points of view facing vast expanses.[12]

There do not appear to be any standard "body space" formats, though certain works, particularly representations focused on details, appear to directly address our bodies by awakening internal sensations. Via a mirror effect, it is possible to feel the physical presence of the other through the trace of a hand on a prehistoric wall[13] or in the sculpted representation of a caress[14]; to perceive the scent of a flower or the taste of a fruit in a still life; or to be caught in painful empathy upon seeing a detail of a tortured body.

Following this survey, there appear to be correlations between "geometries of movement" and the formats and practices of the visual arts, not only in the content of works, the subjects represented, and their actions, but also in the ways the works are shown, placing spectators/viewers in one or another of these "geometries of movement".

---

[7] *"Paris Street; Rainy Day"*, Gustave Caillebotte, 1877.

[8] *L "e Moulin de la Galette"*, Pierre Auguste Renoir, 1876.

[9] *"Louis XIV"*, Louis Petitot, 1836.

[10] *"David"*, Michelangelo, 1501–1504.

[11] *"Map of Hell"*, Sandro Botticelli, 1485.

[12] *"Le plan voisin"*, Le Corbusier, 1925.

[13] *"Chauvet-Pont-d'Arc"*, Cave 36,000 BP.

[14] *"The Cathédrale"*, Auguste Rodin, 1908.

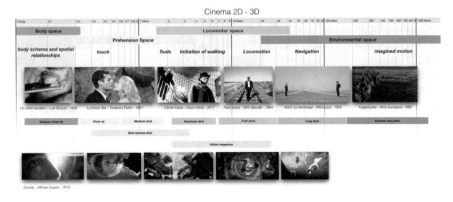

Fig. 12.4 2D-3D cinema

We based our work mainly on examples from classical Western painting and sculpture. It would be very interesting to open this discussion to African and Asian arts, as well as to prehistoric art forms.

In Traditional and Stereoscopic Cinema

Since the advent of film and the invention of cinema at the end of the nineteenth century, artists have been producing two-dimensional and temporal works, moving images that articulate successive points of view and actions in time. This paradigm is of a different nature from the visual arts in terms of forms of display. In the cinema screening room, the images are immaterial (light) and follow each other within the fixed frame of the screen. Spectators are static, not moving, but they are positioned subjectively in the different points of view of the filmed places by the use of different types of shot (Fig. 12.4).

Because the proportions and size of the images are fixed by the screen and the projection distance, the type of shot in cinema is defined according to the content and the presence of the actor in the image. Filmmakers speak of "extreme close-up" or "close-up", framing the eyes or face; the "medium close shot" when the actor is filmed from the torso, waist, or knees up; the "full shot" when the feet are included; and the "long shot" when the character becomes secondary to the environment.[15] As with painting formats, it is possible to find correlations between this scale of framing and the various "geometries of movement".

But cinema cannot be reduced to frame: the other key element of film language is editing, the articulation of the images and actions in time. Types of editing were developed to address specific types of actions. For example, the principle of "shot/reverse shot" editing alternates egocentric and heterocentric "medium shots" of actors in static positions, especially during dialogue or the manipulation of objects. It is used in the distances and for the gestures characteristic of "prehension space". The cinema provides similar editing responses to actions of touch and gaze, as well as in dialogue.

---

[15] Arijon (1983).

Might gaze and speech be seen as proprioceptive "tools" that make it possible to "touch", to test our immediate environment, and if so, could such actions be modelled as cognitive geometries?

In the "locomotor space" zone, another form of editing, "cutting on the movement", is used, in particular in chase or fight scenes. It edits the action in "medium shots" and "long shots" and often includes subjective tracking shots, directly placing viewers in the point of view of the moving character.

Close-ups and establishing shots are used as inserts to punctuate the action or the dialogue, either to focus our attention on an emotion or on what the actors are feeling (body space), or to position actors in their environment (environmental space).

With stereoscopy, a simulation of binocular vision, cinema integrates an additional spatial dimension. Viewers are invited to participate in a novel perceptual experience that immerses them in a spatial point of view. They no longer perceive a flat image, but objects distributed in front of and behind the screen, the proximity of which they can physically evaluate. When viewers watch a stereoscopic film, they are more strongly mobilized corporally than in traditional cinema.[16] It is thus not uncommon in 3D cinemas to see viewers reach out to "touch" shapes or faces positioned closest to them (prehension space), or, though sitting in their seats, to contract their muscles during action sequences (locomotor space).

We also see a change in editing practices in stereoscopy, expressed by the use of longer shots and more frequent use of subjective tracking shots. Viewers need more time to immerse themselves in the space, to feel the presence of actors or the deployment of a movement.

In a previous article,[17] I examined the changing use of the camera by directors when they film in stereoscopy. I studied how, in stereoscopy, the notion of distance to the subject takes precedence over the notion of frame that codifies editing in traditional cinema.[18] I thus proposed, when defining types of shots in stereoscopy, to substitute for the classic framing values (close-up, medium shot, full shot, etc.) a different scale, based on the proxemics zones proposed by Edward T. Hall,[19] that would define shots according to the distance to the subject. The proposal of structuring the environment in "space-time geometries", based on action-perception, appears to be a pertinent new approach for exploring this subject.

In Virtual Worlds, Virtual Reality and Augmented Reality

In recent years, thanks to the development of digital communication technologies, new spatial media are emerging. They allow users, among others communities of artists, to express themselves fully in three-dimensional and temporal spaces by providing the opportunity to act and interact within these digital environments.[20]

---

[16]Okada et al. (2000).

[17]Garnier (2018).

[18]Cutting (1997).

[19]Hall (1971).

[20]Grau (2004).

Virtual World - Virtual Reality

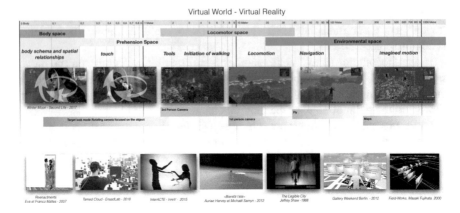

Fig. 12.5 Virtual world—virtual reality

Two main types of practices are currently developing, virtual worlds and virtual reality (Fig. 12.5).

Virtual worlds have mainly developed in the field of gaming and "serious games". They allow a community of players to move and interact with each other in digital spaces. Players, usually seated in front of a screen, are represented in the virtual world by an avatar they can see and manipulate from a computer or game console via a mouse or joystick. Some virtual worlds, such as *Second Life*, have no predefined scenario. Players are free to create their avatars, the environments they will move around in, and the relationships they will develop with other players. These worlds are used above all as spaces for social interaction, where users develop relational, commercial, playful, and artistic relationships over long periods of time.[21] Because these applications are not dedicated to specific tasks imposed by the scenario, the digital interface offers users a lot of freedom of action, in particular in the choice of modes of controlling their point of view in the virtual world. Users are free to choose modes based on their actions and the social context they are in.[22] There are several camera control modes: the camera in the subjective position of the avatar (first player mode), the camera following the avatar (third person mode), and a rotating camera focused on the object being pursued (target look mode).

It has been demonstrated that users project themselves mentally in their representations (phenomenon of embodiment) and become one with their avatar.[23] They then conceive of the virtual environment not from their physical body positioned in front of the screen, but from their avatar immersed in that space.[24] To do that, they can use several modes for controlling the virtual camera, and, unlike in cinema, it is not the director who frames the actions, but the users themselves.

---

[21] Pioneering artists have invested these spaces and developed experiments there, including Chris Marker, Fred Forest, and Yann Ming.

[22] Salamin et al. (2006).

[23] Slater (2009).

[24] Heeter (1992).

We propose to examine how users employ their camera in these situations, with regard to the different "space-time geometries". In general, when the avatar is static and users want to look at it or look carefully at another avatar or object nearby, in either "body space" or "prehension space", they most often employ the rotating camera (target look mode). Users turn the camera onto the observed avatar or object and turn it in an orbit around that point, from a short and variable distance. Thanks to this rotational movement around the object, users can apprehend its volume and location by parallax. It should be noted that in this mode, when users look at an object or avatar, their gaze is oriented toward the object, indicating to other players what their center of interest is. In the same way, when they manipulate or move an object, the outstretched arm signals the action of touching.

In situations of interactive group dynamics (locomotor space), the camera control mode most often used is "third person" mode. The camera is linked to an avatar and positioned behind its back. It frames the avatar in the lower part of the screen, revealing the space facing it. It is thus easy for players to locate themselves in the environment and in relationship to other players.

When players move a long way, especially in quests or pursuits, they favor another camera type, the "first player" mode, in which the camera functions from the eyes of the avatar in subjective vision.

Finally, when they have to cover long distances (environment space), players favor "fly" mode, which allows them to glide above obstacles and provides an unobstructed view of the landscape from a 45° angle, similar to what we would have looking down at a scale model. For very great distances out of visual range, players have recourse to maps on which they can teleport with a click of the mouse.

Virtual reality and augmented reality emerged with the development of visual, haptic, and dynamic interfaces that directly involve the bodies and movements of players in the process of perception of and interaction with virtual spaces.[25] These technologies involve the player's body in a delimited physical space. The virtual space is precisely calibrated so that it is superimposed perfectly on the physical space. The physical space and virtual space thus offer an identical perimeter of action and interaction.

There is a great diversity of virtual reality installations, each of which can be considered as a form of display that specifically favors certain types of actions or perceptions. Here again it is possible to find strategies that seem to favor the expression of one or the other of the "geometries of movement". We will cite a few examples among the great diversity of experiments already carried out:

Works exploring "body space" include systems working with touch or the near presence of another person, such as performances carried out in virtual worlds in which players hesitate to provoke contact between their virtual body and that of another avatar;[26] and, in virtual reality, "The Machine to Be Another",[27] in which

---

[25]Philippe (2018).

[26] "Reenactments", Eva and Franco Mattes, 2007.

[27] "The Machine to Be Another", BeAnotherLab, 2014.

two users exchange their points of view using their headsets and explore the otherness of their bodies.

For "prehension space", there are experiments such as *"InterACTE"*,[28] in which users try to enter into gestural communication with an avatar provided with artificial intelligence; *"The Enemy""*[29] which places us face to face with the avatars of the most dangerous terrorists; and *"Tamed Cloud"*,[30] which allows users to re-appropriate digitized data through gesture and speech.

For "locomotor space" we see the pioneering installation by Jeffrey Shaw, *"The Legible City"*,[31] a bicycle tour of a virtual city; numerous games based on quests and encounters, often warlike but sometimes very poetic, like *"Bientôt l'été"*;[32] and VRroom experiences, virtual reality rooms in which several users explore virtual environments, like the *"ScanPyramids VR"*[33] experience that offers a tour of the pyramid of Khufu.

Finally, for "environment space", we find undertakings that place users in flight position above landscapes or maps, such as *"Birdly"*[34] and *"Devenez avatar"*,[35] using spatial mapping overflights; and in augmented reality we see mapping of personal environments, such as *"Field-Works"*, by Masaki Fujihata[36].

12.4 Discussion

This article has no ambition beyond that of creating a non-exhaustive inventory in order to stimulate thinking about this subject: the simple fact that artists have represented the actions of touching in many different media obviously does not demonstrate the existence of a "geometrie of movement" of "prehension space".

However, this first exploration allows us to glimpse remarkable similarities between this theory from cognitive science and mathematics and various practices in the visual arts and communication: in the correspondences between represented actions and how they are shown; in the codification of "formats"; and in certain practices in which the characteristics of different "geometries of movement" are expressed, notably in their capacity to stimulate changes of systems of reference.

"Geometries of movement"are a new possible theoretical approach that opens research perspectives about the understanding of both perceptive and creative processes. That research will have to be based, on the one hand, on a more finely-tuned

---

[28] *"InterACTE"*, InreV, 2015.

[29] *"The Enemy"*, Karim Ben Khelifa, 2017..

[30] *"Tamed Cloud"*, EnsadLab, 2018.

[31] *"The Legible City"*, Jeffrey Shaw, 1988.

[32] *"Bientôt l'été"*, Auriae Harvey and Michaël Samyn, 2012.

[33] *"ScanPyramids VR"*, Musée de l'architecture de Paris, 2011.

[34] *"Birdly"*, Institut für Designforschung, Zürcher Hochschule der Künste (ZHdK) , 2014.

[35] *"Devenez avatar"*, Amato, Perény, Berthoz, 2015.

[36] *"Field-Works"*, Masaki Fujihata, 2000.

proposition about the mathematical and cognitive characteristics of the different geometries, and, on the other hand, on studies based on large bodies of works that are not limited to the visual arts but are crossed with research in the performing arts and in architecture. It should also employ statistical analyses, case studies, and the realization of experimental paradigms.

Some specific lines of thought can already be proposed:

Despite constant questioning by artists, the notion of format has persisted in art for centuries and finds equivalents in all of the visual arts. One hypothesis would be that the resilience of these formats can be explained by how they correspond to and stimulate the various distances of expression of the "geometries of movement".

With regard to its use in cinema, in particular in shot/reverse shot editing, we think it would be interesting to study whether gaze and speech might be considered as proprioceptive "tools" making it possible to test our nearby environment, particularly in "prehension space".

Finally, it seems to us that the concept of "geometries of movement" is fully expressed in practices related to new forms of spatial communication (virtual worlds and virtual reality). Following this symposium, which is strongly rooted in a trans-disciplinary arts and sciences approach, we feel it is necessary to pursue this dialogue among mathematicians and researchers in the cognitive sciences and in art. Such an approach could lead to the development of study paradigms in virtual reality aiming to better understand the cognitive processes of perception, as well as the processes of artistic mediation.

Acknowledgements I warmly thank Tamar Flash and Alain Berthoz, as well as Daniel Bennequin for accepting me in this deep and dynamic dialogue between reason and emotion, art and science.

References

Arijon D. (1983). *Grammaire du langage filmé*, édition Dujarric, 1983 ISBN: 2-85947-065-4.

Béguin, A. (2001). In A. Béguin (Ed.), *Dictionnaire technique de la peinture: Pour les arts, le bâtiment et l'industrie, volume 1, articles "châssis" and "organisation des surfaces"*, (p. 263).

Bennequin, D., Fuchs, R., Berthoz, A., & Flash, T. (2009). Movement timing and invariance arise from several geometries. *PLOS Computational Biology, 5*(7), e1000426. https://doi.org/10.1371/journal.pcbi.1000426.

Bennequin, D., Berthoz, A. (2017). Several geometries for movements generations. J. -P. Laumond et al. (Eds.), *Geometric and numerical foundations of movements, Springer tracts in advanced robotics* (Vol. 117). https://doi.org/10.1007/978-3-319-51547-2_2.

Cutting, J. E. (1997). *Behavior Research Methods, Instruments, & Computers, 29,* 27.

Garnier, F. (2018). *Un cinéma qui touche : vers un cinéma relief d'auteur, de l'action au sensible.* in *"Stéréoscopie et Illusion"*, published by Presses Universitaires du Septentrion (ArtsH2H). ISBN: 9782757420706.

Grau, O. (2004). *Virtual art: From illusion to immersion.* The MIT Press.

Heeter, C. (1992). Being there: The subjective experience of presence. *Presence: Teleoperators and Virtual Environments, 1*(2), 262–271. xxvii, 25, 30, 31, 32, 34.

Hall, E. T. (1971). *La Dimension cachée (The Hidden Dimension).* Seuil.

Okada, T., Inui, T., Tanaka, S., Nishizawa, S., & Konisi, J. (2000) Functional magnetic resonance imaging of human cognitive processes. *Japanese Psychological Research. Special Issue: Brain Images and Cognitive Functions, 42*(1), 26–35.

Philippe, F. (2018) Théorie de la réalité virtuelle, les véritables usages. Presses de l'Ecole des Mines.

Salamin, P., Vexo, F., & Thalmann, D. (2006). The benefits of third-person perspective in virtual and augmented reality? In *Proceedings of the ACM Symposium on Virtual Reality Software and Technology*-VRST '06 (pp. 27–30). New York, NY, USA: ACM. xxvii, 48, 49, 50, 89.

Slater, M. (2009). Place illusion and plausibility can lead to realistic behaviour in immersive virtual environments. *Philosophical Transactions of the Royal Society of London B: Biological Sciences, 364*(1535), 3549–3557. 34, 55.

Chapter 13
Interaction Between Spectator and Virtual Actor Through Movement: From Child Gestures to Interactive Digital Creation

Marie-Hélène Tramus and Dominique Boutet

Abstract This article is based on the CIGALE project, an interdisciplinary research between digital art, linguistics, and theater on motion capture and interaction with artistic, co-speech and expressive gestures. The aim of this project is to explore an interactive gestural dialogue between spectator and virtual actor. We will explain how our artistic modalities deal with space, time, and movement in a particular way to achieve our specific artistic goal: to encourage the emergence of a gestural interaction between human and virtual actors, thus giving the feeling of a "living" dialogue opening on an aesthetics of improvisation. First, we will describe the design and development process of *InterACTE*, an artistic installation for improvised gestural interaction between a spectator and a digital partner (virtual actor). Then, from seven recorded videos of interactions between the virtual actor and the spectator, we will present the interdisciplinary study we conducted with an artistic approach, and a linguistic approach.

M.-H. Tramus (✉)
University of Paris 8, INREV-AIAC, Saint-Denis, France
e-mail: mh.tramus@gmail.com

D. Boutet
University of Rouen, Dylis, Rouen, France
e-mail: dominique_jean.boutet@orange.fr

T. Flash and A. Berthoz (eds.), *Space-Time Geometries for Motion and Perception in the Brain and the Arts*, Lecture Notes in Morphogenesis,
https://doi.org/10.1007/978-3-030-57227-3_13

13.1 *InterACTE*: An Artistic Installation of Gestural Interaction

Within the framework of the CIGALE project[1] (Tramus et al. 2018), the artistic installation *InterACTE*[2] has been developed using a multi-agent interaction platform that makes it possible to trigger different gestural behaviors of the virtual actor (VA) based on a geometric and kinematic analysis of the movement of both the spectator and the virtual actor.

13.1.1 *Virtual Behaviors Modeled from Gestural Improvisation Between Two Human Actors*

The behaviors exhibited by the virtual actor were based on the analysis of gestural experiments carried out by two humans who played the roles of the virtual actor and the spectator. This revealed recurrences that we modelled according to four behaviors for the virtual actor. Thus, the virtual actor can interact, either by imitating the gestures of the spectator thanks to real-time motion capture of them, by performing a gesture generated in real time by a genetic algorithm, by performing a gesture from the mocap database of recorded human gestures (coverbal gestures, poet gestures in sign language, mime gestures, choirmaster gestures), or by remaining in a rest posture while waiting for the action of the spectator.

13.1.2 *A Gestural Interaction Based on the Geometric and Kinematic Analysis of the Movement of the Human and Virtual Actor*

To develop the multi-agent interaction platform CIGALE (Batras et al. 2016), a finite state machine (FSM) was used to trigger the succession of different behaviors during the actors' gestural dialogue. The gestural activity of both actors was modeled as "perceptions". The virtual actor perceives its own activity and the activity of the spectator, obtained by a continuous analysis of gestures in real time according

---

[1] The CIGALE project (Capture and Interaction with Artistic, Language and Expressive Gestures), supported by the Labex ARTS-H2H and led by the Digital Image and Virtual Reality team of the AI-AC laboratory of the University of Paris 8, the Laboratory Structures Formelles du Langage of the CNRS and the University of Paris 8, UQAM University, the Conservatoire National Supérieur d'Art Dramatique, the Laboratory for Movement Analysis, the Solidanim company and the association Arts resonances.

[2] *InterACTE* (2015) Dimitrios Batras, Judith Guez, Jean-François Jego, Marie-Hélène Tramus, University Paris 8 Campus Exhibition, Ars Electronica 2015. See the video: https://www.youtube.com/watch?v=U-Kqv-xeI3k.

to their positions, their orientations, and their kinematics. A comparison of both "perceptions" causes the transition from one behavior to another.

Several "perceptions" were taken into account for the triggering of behaviors. The orientation of the spectator's head or torso is used to indicate whether the spectator is present and ready to establish a dialogue with the virtual actor. The positions of the virtual actor's hands in relation to its sagittal plane are compared with those of the spectator. The purpose is to estimate where the latter produces a gesture in space so that the virtual actor's behavior occurs in the same place. The kinematics analysis of hand gestures provides the successive derivatives: *velocity, acceleration* and *jerk. Velocity* is used to assess the overall activity, *acceleration* as an estimation of change of the overall activity and as an indication of phase synchronization, and *jerk* is used to detect a change in acceleration, which can match either a pause or a discontinuity in the movement. The acceleration or jerk values, depending whether they are low, medium, or high, are associated with different transitions of the virtual actor's behaviors.

13.1.3   Two Configurations in the Real and Virtual Spaces

Two configurations of the installation have been implemented in real space and into the virtual space. The first is based on the two-dimensional projection onto a screen (Fig. 13.1), seen as an interactive shadow theater composed of the real shadow of the spectator and the digital shadow of the virtual actor. Interaction is made possible via a Kinect motion sensor. In this case, the shadow of the virtual actor is projected onto the screen. The virtual actor itself is not visible to the spectator. The VA becomes

Fig. 13.1 Shadow theater

Fig. 13.2 Virtual reality

visible only during the second part of the interaction (Fig. 13.2), when the spectator wears a virtual reality HMD helmet (Head Mounted Display).

In the second configuration, thanks to the HMD helmet, the spectator is immersed in another world entirely made of virtual reality (Fig. 13.2). In this three dimensional virtual world, the spectator faces the virtual actor, perceived in relief and with which the spectator can interact using gestures within the entire 3D space. Interactions are achieved by the Kinect motion sensor.

13.2 Analysis of Video Recordings of Gestural Interaction Between Spectator and Virtual Actor

From seven video recordings of interactions[3] between the virtual actor and the spectator we examine what promotes the continuity of a gestural dialogue, and further, what gives the feeling of a "living" improvisation. For this study, we developed an interdisciplinary approach by jointly carrying out an analysis from an artistic point of view (analysis of the forms of dialogue emerging from the gestural interactivity) and a linguistic analysis (dyadic communication, symbolic gestures generated in the arms, the pronosupination of toddlerhood).

---

[3]Video recordings made during two exhibitions: University Paris 8's Campus Exhibition of the Ars Electronica International Festival, Linz Austria, September 2015; Les Vitrines du Labex Arts-H2H, Theater Gérard Philipe, Saint-Denis, France, October 2015.

13.2.1 Analysis from an Artistic Point of View

In interactive digital art, the interactivity between the artwork and the spectator is the place where an aesthetic of relationship emerge (Tramus 2017). With this approach, the artistic analysis focuses on the interactivity between the virtual actor and the spectator. The analysis of interactivity requires bringing out the diversity of these forms of relationships.

13.2.1.1 Imperfect Imitation of the Gesture

The gestural imitation performed by the virtual actor creates a link with the spectator. This imitation exceeds that of a simple mirror because it is imperfect and it is done with a time delay. Thus, it might give the impression that the virtual actor tries, with clumsiness and hesitation, to follow the spectator's gestures, as if it were "alive". The spectator has the sensation of being imitated by the virtual actor which would provoke in him/her a sensorimotor resonance with the artificial entity opening perhaps on a form of empathy. This initiates an approximate synchronization between the movements of the two actors, both in their forms and dynamics, and triggers a harmonious dialogue.

13.2.1.2 Alternation Between *Imitation* Gesture and *Idiosyncratic* Gesture

Suddenly, the imitation is broken by the occurrence of idiosyncratic gestures by the virtual actor, either drawn from its mocap database (natural movements) or generated by a genetic algorithm (artificial movements). These gestures have nothing to do with the form of the spectator's movements, yet corresponding to an almost similar kinematic. This break tends to revive the dialogue by prompting surprise and giving the illusion of a certain autonomy of the virtual actor. The spectator is puzzled and tries to understand the unexpected gestural response of the VA. We also see that, in turn, the spectator is influenced by the virtual actor's gesture and begins to imitate it. Then, the virtual actor initiates a new sequence of gestural imitation, and so on.

13.2.1.3 Emergence of Varied and Improvised Dialogues During Interaction

Varied and improvised dialogues emerge during interaction (Fig. 13.3). Each spectator performs his/her proper gestures according to his/her own attitudes and also projects a kind of scenario inducing a specific interaction strategy. Nevertheless, three principal types of dialogue strategies can be identified: exploration, encounter, and the search for aesthetic harmony.

Fig. 13.3 Varied dialogues

Exploration Strategy[4]

This spectator strategy explores the behaviors and reactions of the virtual actor in order to figure out ways to interact with it. This strategy appears mostly at the beginning of interaction with the virtual actor in several experiments. They test the virtual actor's abilities by making all sorts of movements to find out what movements the virtual actor is able to perform, such as moving its arms, chest, or head; and conversely, to discover what it cannot do, like moving fingers, bending down, etc. Some spectators continue this practice throughout the whole exchange. Hence, these exploratory games themselves become the core of the dialogue between the two actors, a kind of body competition.

Encounter Strategy[5]

Some spectators consider the gestural dialogue in the form of an encounter. They greet the virtual actor with their hands, or curtsy, or open their arms to welcome him, etc. Other approach, caress its face, make a farewell gesture and leave, etc. Here again, dialogues emerge from the interaction between the spectator and the virtual actor. The VA reacts to these gestures, either by responding to the spectator in its imperfect imitation,—by sketching a salute, or opening the arms clumsily, etc.—or by an idiosyncratic gesture. In the latter case, the spectator leave the usual dialogue of communication associated with the encounter to start an unusual dialogue with a strange entity.

---

[4]See the video: https://www.dailymotion.com/video/x71d509.

[5]See the video: https://www.dailymotion.com/video/x71d6dl.

Fig. 13.4 Harmonious dialogue

In one video sequence, the spectator tries to touch the shadow of the virtual actor's hand with his own hand, as if to shake hands to say "hello". But, this fails because the virtual actor's hand seems to avoid the spectator's. At first, a dialogue appears in the form of a rather lively chasing game. Then, the spectator slows down the movements of his hand, which allows him to approach the hand of the virtual actor. Thus appears a dialogue in which the hands slowly draw a sort of choreography in space.

Aesthetic Harmony Strategy[6]

Some spectators try to bring out an aesthetic harmony in the dialogue by performing simple and slow movements in order to avoid losing the thread of gestural interdependence with the virtual actor. In one sequence for example, a woman pays attention to the imitation of the virtual actor's gestures. She seeks harmony by developing a very slow and continuous gesture that she constantly adapts according to the movements of the virtual actor. She enters then into resonance with the virtual actor by putting herself in the same rhythm. This causes subtle and delicate gestural interactions: the hands of the two actors follow each other, touch each other, and move together to create a kind of harmonious choreography (Fig. 13.4). Then, a break appears, reactivating the dialogue which carries on by movements resulting from hesitant and shared mimicry.

Beyond gestures with a goal of exploration or significant gestures such as welcoming, a gesturality appears that seems free of meaning. This situation explores new artistic forms of gestural dialogue. This kind of interaction opens up an aesthetic reflection on the forms of dialogue between spectator and virtual actor.

---

[6]See the video: https://www.dailymotion.com/video/x71d6wr.

13.2.1.4 Playing with the Laws of Natural Movement

For artists designing interactive virtual actors in relation to spectators, research on the brain's control of movement and perception is of great interest.

In *Le Sens du mouvement* (Berthoz 1997, p. 159), Alain Berthoz addresses, among other things, the question of reducing the number of degrees of freedom and the laws simplifying natural human movement. Among these laws, Berthoz points out that the two-thirds power law[7] (Lacquantini et al. 1983; Viviani and Schneider 1991; Viviani and Flash 1995) linking kinematics (velocity) to geometry (curvature) would make it possible to simplify the control of movement. As an illustration of this law, he gives the example of the movement made by the hand to draw an ellipse on a sheet of paper with a pencil. This example shows that the velocity varies along the trajectory; it is faster where the curvature is weak and slower where the curvature is strong.

Alain Berthoz also points out, in *La Simplexité* (Berthoz 2009, p. 110), that if the movement of drawing is performed in three-dimensional space, the movements of the hand and arm are more complex. They obey the 1/6th power law[8] linking velocity not only to curvature, but also to torsion (Maoz et al. 2009; Pollick et al. 2009; Maoz and Flash 2014). For him, this law can be derived from the fact that the brain itself functions in non-Euclidean geometry. Thus, for researchers who have mathematically modeled natural human movements, such as Tamar Flash, the generation of natural movements by the brain are based on a mixture of several geometries: Euclidean and other non-Euclidean geometries, for example, affine and equi-affine geometry (Flash and Handsel 2007; Bennequin et al. 2009).

Alain Berthoz also insists on another important discovery (Berthoz 2009, p. 110): "the two-thirds power law also underlies the perception of natural movement" (Viviani and Stucchi 1992). This leads him to argue "that if roboticists, or digital image specialists, want to ensure that artificial characters (avatars or humanoids) are perceived by the spectator as living creatures, or at least as close as possible to living ones, their movements must obey this law" (Berthoz 2009, p. 110).

In the installation InterACTE, the gestures of the virtual actor during the dialogue with the spectator come mostly from motion captures, either in real time from the spectator's gestures, or pre-recorded from the gestures of different human actors (mime, sign language poet, choir director and linguist). In this case, the spectator perceives the virtual actor's gestures as natural, responding to the laws of natural movement since they come from captured human gestures. This seems to enhance gestural dialogue, because the spectator then perceives the gestures of the virtual actor as if they were made by the same sort of living entity.

But from time to time, the virtual actor's gestures are generated by a genetic algorithm. Initially, they do not respond to the laws of natural motion, because they are random. Then they evolve progressively, converging with a captured human gesture, representing the law of adaptation. In the course of this evolution, the gestures are imperfect, the result of hybridization between randomness and the laws of natural

---

[7]$A = KC^{2/3}$ with A: angular velocity and C: curvature.

[8]$v = \propto K^{1/3}.|\tau|^{1/6}$ with v: tangential velocity; K: curvature; τ: torsion; $\propto$: coefficient.

movement, and thus open the way for artistic creation. In this case, the spectator perceives unfamiliar movements, which are both surprising and disconcerting, but which can also stimulate the imagination.

This artistic game of playing with the laws of natural movement, respected or not, or even hybridized, seems conducive to the spectator questioning the nature of the gestural interaction had with the artificial entity, as well as the nature of communication through gesture in general. Above all, this game promotes the invention of unusual, improvised gestural dialogues.

13.2.2 Linguistic Analysis of the Spectators' Gestures

13.2.2.1 Dyadic Communication Is Anchored in the Arms

Symbolic Gestures

Before the presentation of the linguistic study, it is important to qualify the semiotics of gestures according to body segments and especially concerning one kind of gesture: the gestures generated in the arms. For the purposes of this study, we make the assumption that these gestures represent a dyadic form of communication: the communicative functions are oriented to the relationship with the interlocutor (Boutet 2015, p. 126).

This type of gesture is illustrated in Rigaud's 1701 portrait of Louis XIV, a portrait of social and political significance (Fig. 13.5).

The position of the arm akimbo opens the angle between the arm and the side of the torso (*abduction* of the arm). The common posture of superheroes—hands on hips—is not different. The abduction of the arms has historically expressed power in and of itself.

Contrarily, an *adduction* of the arm (elbows close to the body), generally expresses submission. A great curtsy, such as the one Jack Nicholson makes in Fig. 13.6, or a military salute both express a form of submission.

In other words, the antonyms—power and submission—are often expressed by this degree of freedom (DOF) of the arm: the abduction/adduction.

Flexion and extension demonstrate another degree of freedom. Flexion often expresses a request, even if the flexion is complete (arms up). Other examples of flexion include postures of imploration (the classical representation of an intercession), and sometimes are a mix between a request and an assertion («please don't shoot» and «we are unarmed»). Examples of this can be seen in the recent Black Lives Matter demonstrations in the USA.

Extension postures, by comparison, place the elbows behind the torso and indicate sacrifice or constraint. This is exemplified in images of various constraints or torture devices, such as handcuffs or gallows.

Fig. 13.5 Louis XIV Sacre position akimbo, Hyacinthe Rigaud, 1701, Musée du Louvre. Left arm in abduction (akimbo)

Interestingly, an expressionist attitude produced in movies during the 20 and 30's symbolizing sacrifice put the chest and the head forward with the arms behind the body. The extension of the arm can be glossed as a self-sacrifice attitude or a self-constraint attitude.

To summarize, these two types of degrees of freedom, abduction/adduction and flexion/extension are associated with dyadic communication. In terms of human relationships, they express antonymic meanings: abduction/flexion expresses power and request, while adduction/extension expresses submission and self-sacrifice. We can see a coupling between the meaning and the forms: the antonyms are formally associated with the poles of the arms. The meaning carried by these poles responds to a polar geometry in which the amplitudes increase each meaning.

Pronosupination in Toddlerhood

We have discussed the gestural semantics corresponding to degrees of freedom in the shoulder. Now, we work our way further down the upper limb with this degree

Fig. 13.6 Jack Nicholson performing a great curtsy. Right arm in adduction

of freedom (DOF) made available in the partnership of the forearm and hand and required for pronosupination, or the rotation of hands facing up and down. The absence of this DOF during toddlerhood (until 36 months) is especially interesting. The maturation of motor schema and of the motor control carry out in distalization (Konczak et al. 1995). The DOF expressed in pronosupine gestures is one of the last to be developed. The authors of this article have uncovered no study showing active pronosupine range of motion in toddlerhood. This suggests that the movement is not established during this age.

In French nursery rhymes[9], one of the purposes of the associated pantomime is to trigger pronosupine movements by imitation. These examples are in French, but it seems likely that you can find the same gestures in other languages. At the end of these clips, you will see a toddler trying to make a pronosupine movement. The difficulties are apparent for him as he simulates this supination/pronation movement using an abduction/adduction of the wrist in place of complete extension. He is unable to perform a proper pronosupine movement.

---

[9]See the video: https://www.dailymotion.com/video/x72f1q5.

Our software imparts the same behavior. Pronosupination is not perceived by this Kinect 1 device. Therefore, the virtual actor is not able to perform pronosupination, just like a toddler. In a certain way, the spectator is in front of toddler behavior when interacting with the VA. This configuration invites us to consider the interaction as a dyad between an adult (spectator) and a toddler (VA).

Two Hypotheses

We make two kinds of assumptions. The first one concerns the impact on gesture caused by this interaction. The second is about the consequences on the interaction.

Hypothesis 1 expects a quick gestural adjustment from the spectator vis-à-vis the gestural constraints of the VA and especially due to its inability to detect or perform pronosupination.

Hypothesis 2 expects that the amount of arm movement at the end of the interaction should correlate with the report of this amount and should have an impact on the dyadic communication. To summarize the reasons for this hypothesis, a software choice for the Mocap device could affect the terms of the relationship between human and virtual actor and trigger expected behaviors when certain conditions are brought together.

Coding the Gestures

The gestures of the interaction are without any clear or emblematic meanings. They are closer to dance or a series of improvised movements.

Seven corpora[10] have been coded and studied. For each corpus, we coded two parts. The first part is the ten first seconds of the interaction with the VA shadow. The second part is taken from the last ten seconds. We have measured the most opposite extremes of each interaction in order to clarify the impacts of this kind of interaction on spectator's gestures.

The mean length of each corpus is about 1 min. So if there is an impact, the effects are quite quick. We coded all of the movements according to each degree of freedom, from the hand to the shoulder, using ELAN software (Sloetjes and Wittenburg 2008). Each unit of coding corresponds to a cusp of the amplitude or extent of a movement, or a plateau value. Effectively, each transcribed unit corresponds to a position in the range of movement (expressed by degree). The duration of each unit depends on the movement, or of the maintained posture for each degree of freedom. All the results have been extracted on an excel sheet.

---

[10]Corpora, in this context, refer to the tools used in corpus linguistics. A corpus is a unit of linguistic data (an annotated recording, a sample taken from real world examples of natural movement and behavior patterns) that represents body motion. Corpora require specialized software tools to be viewed and analyzed.

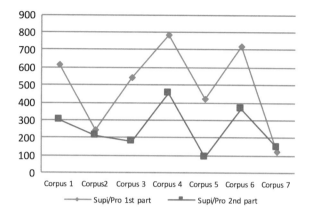

Fig. 13.7 Variations in the extent of pronosupine motion between the first and second 10 s of interaction. The Y-axis, expressed in degrees, represents the cumulated angles of pronosupine gestures produced by the spectator. The X-axis represents each annotated recording

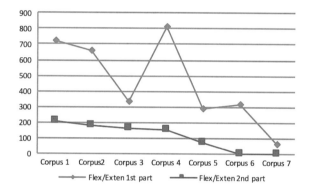

Fig. 13.8 Variations in the extent of flexion/extension between the first and second 10 s of interaction. The Y-axis represents the cumulated range of flex/exten gestures produced by the spectator in degrees

Results

According to Hypothesis 1, we should have a decrease in the cumulated range of motion of pronosupine movement between the first part and the second part for each corpus due to the deprivation of pronosupination input and motion capabilities of the VA.

In the chart (Fig. 13.7), the blue line represents the cumulated range of motion of pronosupination in the first ten seconds. The red represents the cumulated range of motion of the same movement, but at the end of the interaction (2nd ten seconds). In six cases out of seven, the cumulated angle measurements of pronosupine movement diminish at the end of the interaction. Hypothesis 1, therefore, appears to be confirmed.

In a sub-hypothesis (1.2), the flexion/extension of the hand should decrease in the second part of each corpus. Here again (Fig. 13.8), the blue line represents the range of motion of flexion/extension during the first part of interaction; the red, the final part of interaction.

The impact of the lack of pronosupine movement available to the Kinect sensor correlates to the flexion/extension of the spectator. The decrease in range of motion

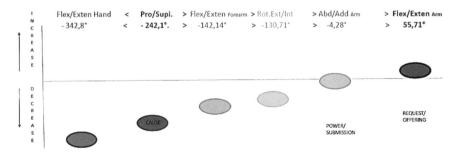

| | Flex/Exten Hand | < | Pro/Supi. | > Flex/Exten Forearm | > Rot.Ext/Int | > Abd/Add Arm | > Flex/Exten Arm |
| | - 342,8° | < | - 242,1°. | > -142,14° | > -130,71° | > -4,28° | > 55,71° |

Fig. 13.9 Variations of the cumulated extent for each degree of freedom between the 1st and 2nd part of the 7th interactions. The absolute increase of the flex/exten of the arm is obvious, but one can not neglect the relative increase of the abd/add of the arm compared the diminution of the forearm and the hand movements

is more pronounced than the decrease in the pronosupine movement. We can see here that the degrees of freedom on the hand are dependent on each other. When the movement of a DOF diminishes, the movement of another DOF is prone to decrease.

What is going on with the forearm and the arm? As we move away from the prono/supine i/the influence of the absence of pronosupine movement diminishes and ii/the range of the movement is moving up along the upper limb, toward the arm. This displacement along the upper limb is not linear. We must take into account the inertia of each segment, the consequences of the diadochal[11] movement (MacConaill 1948; Codman 1984; Cheng 2006), which link the arm's abd/adduction to its exterior/interior rotation and its pronosupination. This fact shortens the distance between the pronation/supination and the abduction/adduction of the arm, despite the inertial distance between the hand and the arm.

Of course, the hand moves (by displacement), following the more numerous movements of the arm in the second part of corpora, but at least it does not move in a proper movement. The gesture of request or offering (flexion/extension of the arm) and power or submission (abduction/adduction of the arm) should have an impact on the projective space of the hands (Rudrauf et al. 2018). Without detailing these, they require marked positions in all three degrees of freedom (pronation/supination, flexion/extension and abduction/adduction), two of which have been shown to have reduced amplitudes in the second step. Deprived of this additional projective space on the hands, the expression of these dyads is not as perceptible as it could be if it were amplified from the arms to the hands.

In short, we can see that the increase affects the arm's flexion/extension, most likely due to kinematics and anatomical reasons (Fig. 13.9). Such increasing of movement switches the spectator's behavior toward dyadic communication made of requests and offering. The evidence in this study supporting such communication merits continued research using a mocap device with IMU (Inertial Measurement Unit) sensors.

---

[11]Diadochal motion refers to two successive movements in any joint where a rotation of the bone is possible, which imposes conjunct rotation upon the bone that has been moved.

Outcomes of This Study

A technical constraint of the Kinect device results in a gestural behavior similar to the gestures of a toddler, who presents the same difficulty in producing pronosupine movement. The effect on a spectator is obvious and almost immediate, resulting in the displacement of the movement to another body segment according to proximity. This outcome emphasizes the arm flexion/extension of the spectator. The mimetic effect is child-like, similar to how we speak to a child. And the symbolic effect implies cooperation, where the spectator expresses request (flexion) and self-constraint (extension) toward the VA, therefore expressing a connection with the virtual actor, to some extent.

This behavior facilitates communication with another through a prompting toward dialogue, a bit like a game of Ping-Pong between the virtual and spectator. Through this game, the virtual actor becomes a partner.

Empathy, as a faculty to perceive the *Other* (VA) as an agent with intention, seems to be reinforced by this expressive and dyadic gestural communication. This study reveals an adaptive behavior from the spectator, who adopts the constraints of motility recognized in the virtual actor, even while the spectator is not conscious of these constraints (i.e. the absence of pronosupine movement).

This study suggests that we could instead constrain the movement of the shoulder and observe the consequences on the gestures of the spectator, for instance by observing the impact on hand motion. What would the impact be on the gestural communication? Shoulder movement constraint is reminiscent of the behavior seen in elderly people having arthritic pain that affects the shoulder more than the wrist.

13.3   Conclusion

Imperfect imitation of the spectators' gestures by the virtual actor, alternation between imitative and idiosyncratic gestures, emergence of varied and improvised dialogues during interaction, game with the laws of natural movement, respected or not, or even hybridized, are all aspects that contribute to the re-launching of dialogue and giving the impression of a "living" dialogue. But more profoundly, we observe that a gesturality of the virtual actor is similar to that of a toddler and would arouse empathy, encouraging a deepening of interaction. Does this not accompany the aesthetic intention in the center of the *InterACT* installation, which aims to establish an improvised gestural dialogue leading to a motor and emotional resonance of the spectator? Doesn't *InterACT*'s choice to set up a situation of gestural improvisation, marking a desire to emphasize spontaneity and giving free rein to creativity, invite us to find ourselves with the gestural games of childhood, with all their playful and aesthetic pleasures?

If, according to Alain Berthoz, "What makes us as human beings unique is precisely the power to separate from this determinism that confines us to a reality tied to our needs and to our senses, by virtue of the remarkable capacity of our minds to execute those vicarious processes with which it is equipped in order to escape reality—or to escape *perceived* reality." (Berthoz and Tramus 2015) (Berthoz 2013), would these interactive artistic installations with a virtual actor be the expression of a creative vicariance?

Acknowledgements We will lean our presentation on the CIGALE project, an interdisciplinary project between digital art, linguistic, and theater *on Motion capture and interaction with artistic, co-speech and expressive gestures,* granted by the Labex Arts H2H (2012–2015). We particularly thank: Ariel Alonso, Dimitrios Batras, Brigitte Baumié, Marion Blondel, Michel Bret, Fanny Catteau, Clara Chabalier, Julie Châteauvert, Chu-Yin Chen, Sébastien Delacroix, Florine Fouquart, Amélie Gabriel, Nefeli Georgakopoulou, Judith Guez, Patrice Guyot, Chen-Wei Hsieh, Jean-François Jégo, Isabelle Lemaux, Sébastien Lenglet, Marie-Thérèse L'Huillier, Julien Lubek, Xavier Maurel, Guillaume Metais, Anne-Marie Parisot, Philipe Pasquier, Isaac Partouche, Ilaria Renna, Thecla Schiphorst, Jean-François Szlapka, Michel Thion, Coralie Vincent.

Thanks to Dominique Boutet, who has just passed away, for all his rich work: research on co-speech gestures, on sign language and on the relationships between artistic, linguistic and expressive gestures. The publication of this article may serve as a tribute to this researcher who has opened an original way of kinesiological analysis that contributes to our understanding of how meaning and expressiveness emerge from human body gestures.

Bibliography

Batras, D., Guez, J., Jego, J. F., & Tramus, M. H. (2016). A virtual reality agent-based platform for improvisation between real and virtual actors using gestures. In Proceeding VRIC '16. Virtual Reality International Conference—Laval Virtual, March 2016, New York, NY, USA ©2016. Paper No. 34. ISBN 978-1-4503-4180-6.

Berthoz, A. (1997). *Le Sens du mouvement*. Paris: Odile Jacob.

Berthoz, A. (2009). *La Simplexité*. Paris: Odile Jacob.

Berthoz, A. (2013). *La Vicariance: le cerveau créateur de mondes*. Paris: Odile Jacob.

Berthoz, A., & Tramus, M. H. (2015). Towards creative vicariance. *Journal Hybrid* 02 l 2015. http://www.hybrid.univ-paris8.fr/lodel/index.php?id=590. Accessed 19 Jan 2020.

Boutet, D. (2015). Conditions formelles d'une analyse de la négation gestuelle. *Vestnik of Moscow State Linguistic University, Discourse as social practice: priorities and prospects, 6*(717), 116–129.

Bennequin, D., Fuchs, R., Berthoz, A., & Flash, T. (2009). Movement timing and invariance arise from several geometries. *PLoS Computational Biology, 5*(7), e1000426.

Cheng, P. L. (2006). Simulation of Codman's paradox reveals a general law of motion. *Journal of Biomechanics, 39*(7), 1201–1207.

Codman, E. A. (1984). The shoulder: Rupture of the supraspinatus tendon and other lesions in or about the subacromial bursa. RE Kreiger.

Flash, T., & Handzel, A. A. (2007). Affine differential geometry analysis of human arm movements. *Biological Cybernetics, 96,* 577–601.

Konczak, J., Borutta, M,. Topka. H., & Dichgans, J. (1995). The development of goal-directed reaching in infants: Hand trajectory formation and joint torque control. *Experimental Brain Research*, 106(1), 156–168. https://doi.org/10.1007/BF00241365. Accessed 19 Jan 2020.

Lacquantini, F., Terzuolo, C., & Viviani, P. (1983). The law relating the kinematic and figural aspects of drawing movements. *Acta Psychologica, 54,* 115–130.

MacConaill, M. A. (1948). The movements of bones and joints. *Journal of Bone & Joint Surgery*, *British Volume, 30-B*(2), 322–326.

Maoz, U., Berthoz, A., & Flash, T. (2009). Complex unconstrained three-dimensional hand movement and constant equi-affine speed. *Journal of Neurophysiology, 101*(2), 1002–1015.

Maoz, U., & Flash, T. (2014). Spatial constant equi-affine speed and motion perception. *Journal of Neurophysiology, 111,* 336–349.

Pollick, F. E., Maoz, U., Handzel, A. A., Giblin, P. J., Sapiro, G., & Flash, T. (2009). Three-dimensional arm movements at constant equi-affine speed. *Cortex, 45,* 325–339.

Rudrauf, D., Bennequin, D., & Williford, K. (2018). The moon illusion explained by the projective consciousness model. arXiv:1809.04414 [q-bio]. http://arxiv.org/abs/1809.04414. Accessed 19 Jan 2020.

Sloetjes, H., & Wittenburg, P. (2008). Annotation by category: ELAN and ISO DCR. In LREC. http://lrec-conf.org/proceedings/lrec2008/pdf/208_paper.pdf. Accessed 19 Jan 2020.

Tramus, M. H., Chen, C. Y., Guez, J., Jego, J. F., Batras, D., Boutet, D., Blondel, M., Catteau, F., Vincent, C. (2018). Interaction gestuelle improvisée avec un acteur virtuel dans un théâtre d'ombres bidimensionnelles ou au sein d'un univers virtuel en relief: l'illusion d'un dialogue? In M. Almiron, E. Jacopin & G. Pisano G (dir.), Stéréoscopie et Illusion (pp. 281–299). Lille: Presses universitaires Septentrion.

Tramus, M. H. (2017). La question de l'émergence à travers l'évolution des arts de la participation: ceux de l'interaction, ceux de l'interactivité numérique, ceux de l'immersion, jusqu'à ceux de l'autonomie. In X. Lambert (dir.) Poïèse/autopoïèse: art et systèmes, Coll. Ouverture philosophique, série Arts vivants, (pp. 137–146). Paris: L'Harmattan.

Viviani, P., & Schneider, R. (1991). A developmental study of the relation between geometry and kinematics of drawing movements. *Journal of Experimental Psychology (Human Perception), 18,* 603–623.

Viviani, P., & Stucchi, N. (1992). Biological movements look uniform: Evidence of motor-perceptual interactions. *Journal of Experimental Psychology: Human Perception and Performance, 18*(3), 603–623.

Viviani, P., & Flash, T. (1995). Minimum-jerk, two-thirds power law, and isochrony: Converging approaches to movement planning. *Journal of Experimental Psychology: Human Perception and Performance, 21,* 32–53.

Printed in the United States
by Baker & Taylor Publisher Services